Tony and the cows : a true story fr

TONY AND THE COWS

TONY AND THE COWS

A True Story From The Range Wars

Will Baker

Confluence Press, Inc.
Lewiston, ID 83501

ACKNOWLEDGMENTS

For their contributions of time and valuable information, I am grateful to Babe Penn, Tom Bill Black, Michael Facciuto, and Colleen Taugher. Special thanks also to Deanie Roush for the timely snapshot that became our book jacket.

Editors Stephen Corey and Janet Wondra helped me sharpen a briefer version of this story for *The Georgia Review*, and Jim Hepworth at Confluence Press gave me encouragement and savvy counsel through three rewrites of the whole. I am much obliged.

—W.B.

Copyright © 2000 by Will Baker.

Publication of all Confluence Press books is made possible, in part, by the generosity of the Idaho Commission on the Arts, Lewis-Clark State College, and Washington State University.

Cover design by Lewis Agrell
Text design by John K. Wilper

FIRST EDITION
10 9 8 7 6 5 4 3 2 1

ISBN: 1-881090-35-3

SF
85.35
. W4
B35
2000

Published by
Confluence Press, Inc.
Lewis-Clark State College
500 Eighth Avenue
Lewiston, ID 83501
(208) 799-2336

Distributed by
Midpoint Trade Books
1263 Southwest Boulevard
Kansas City, KS 66103
(913) 831-2233
Fax (913) 362-7401

CHAPTER ONE

This was the order of human things; first the forests, after that the huts, thence the villages, next the cities, and finally the academies.
—Giambattista Vico, *La Scienza Nuova*

The woman was perhaps in her early thirties and dressed in a style one might call wilderness grunge: painter's pants, trail boots, straw hat, T-shirt over an unrestrained bosom. This particular shirt bore an arresting image and motto: a composite of reproductions of famous photographs of American Indian warriors (Geronimo, Sitting Bull, Joseph et al.), and under this group of dark, unsmiling faces, a blunt caption: My Heroes Have Always Killed Cowboys.

We were on the Trinity River in northern California in 1995, nearing the Fourth of July. Ten weeks ago Timothy McVeigh and his helpers had blown up the Oklahoma City Federal Building, and a week after that the Unabomber's latest device murdered a timber executive. Earlier in our conversation these events had been mourned, chiefly because they made the public apprehensive at any sort of activism, but now the woman in the cowboy-killer shirt and another veteran Earth First!er were talking about memorable actions pulled off in the old days, about the close calls and wild flights, the high humor and sly tricks, who got jailed and who didn't. The reminiscences blended comedy and menace, subtle braggadocio and self-mockery. Like battle and bear stories, they carried an implication that to *really* get it you had to have been there.

Before long the discussion swung to speculation on possible future actions, including those feasible for this gathering. Any such maneuver would obviously involve Palco (the Pacific Lumber Company) in Scotia, where the last of the unprotected giant redwoods were about to roll into the saws. But there were no details. The speculators were circumspect, allusive, noncommittal. The subtext was after all the breaking of certain laws, the taking of certain risks. Everyone understood you had to be ready to go to the wall to protect the ancient forest, home to the threatened marbled murelet and many other native life-forms whose right to exist was at least as venerable as your own. It did not need to be said—would not be cool to say—that this would also be a time for daring, heroism, perhaps martyrdom. A very small band of ecowarriors, armed with songs, placards, and a few volunteer lawyers, would be pitted against a corporation with its bulldozers and trucks and protective sheriffs. The new warriors were tacitly readying themselves, drumming up spirit for a move—modern and nonviolent of course—to take down the tree-killers, just as Geronimo took down the cowboys.

In 1903 the State of Idaho began requiring hunting licenses

(at one dollar each), but uniformed wardens did not appear in Scott Valley, site of my grandfather's homestead, for another twenty years. By then the road in from Cascade was passable to motor vehicles and a checkpoint could be set up. This development was a serious affront to young bucks like my father and his brothers, who had grown up in that valley on a little quarter-section cattle ranch. There (so they claimed to me a generation later), Granddad would drop a salt block thirty yards from the back door and thereafter step out to shoot a fat deer any morning he saw the need.

So Waldo was in a sullen mood when, inevitably, he and my uncle Jess one day had to pull over the wagon and present their new license and a fresh kill to a pair of game wardens, their peers in age and size. One of the wardens grew talkative and observed that they had a nice four-year-old, judging by his rack. No way to judge, Waldo commented; look at the teeth. Fairly common knowledge, the warden went on. Some people know deer, my father added, because they hunt and shoot and eat them, while other people sit on their behinds reading books on the subject. The discussion continued in this vein for a few minutes longer and ended when my father broke the warden's arm.

Uncle Jess often revived this little story at our garrulous and profane family reunions. It was one installment in a miniseries dramatizing rebellion against the state and big business (which sinister powers regularly employed several members of our clan). It was an informal and sporadic, but lifelong, insurrection. At the age of seventy Waldo was fined one hundred dollars for contempt of court because he refused to sit down and terminate his generic Wobbly lecture, provoked upon this occasion by a capitalist extortion plot involving the arbitrary placement of parking zones. At seventy-three, Jess spent a night in jail and absorbed a fine equal to every cent in his pockets for taking a swing at an Arizona patrolman who had the temerity to stop him for some imagined infraction. His wife, Nelly, a bit of a woman who normally laughed at everything and everybody, also had a streak of cussedness with regard to authority figures. With a butcher knife she herded a deputy sheriff off her front porch to discourage him from continuing his search for her informally adopted ward (a fifteen-year-old wetback) who was hiding in the barn. Most of the men in the family accepted such tribulation as a badge of honor, the price of principle, and—as they saw it—a sure sign of how swiftly the country was going to hell in a handbasket.

My father and his brothers did not, however, grow up to be cowboys like their father, who had ridden the Chisholm Trail. The Baker ranch was too small and poor to survive the Depression. Waldo ended up logging for wages most of his life, and Jess repaired Cats and loaders at his shop in town. They were reduced to a house and a lot, dependent on the dollar for their bread, out of work whenever a logging show shut down.

Their generation marked the end of frontier life. No more surviving on venison and trout out of the backyard, or catching a horse and riding in any direction at one's pleasure, or hiking on homemade snowshoes to set traps. They never quite got over the loss, and I think that was why they were so fond of stories in which the ordinary little guy triumphed over the big outfits or at least spit in their eye, and why they were so cantankerous in the presence of uniforms and directive signs, or later, whenever they saw license plates from California. They loved that too-small, poor old ranch in the wild woods, and they felt, even as young fellows, the pressure of forces (market forces, demographics, development, modernization) driving them—and tempting them—inexorably away from that life. And this was surely the reason—though he wouldn't have puzzled it out so far—Waldo broke the warden's arm.

The second day of the 1995 Trinity River Rendezvous, right after receiving my personal five-gallon plastic receptacle for keeping the ancient forest free of human waste, I attended a welcoming ceremony for new arrivals. I call it a ceremony because it took the form of a big circle in a wide meadow and was punctuated by an occasional chorus of wolf howls, a standard Earth First! rallying signal. Otherwise it was a rather informal affair, with a good deal of banter and laughter in and around the introductions, announcements, and open discussions.

It was the first formatted encounter between veterans and newcomers, some of whom had crossed the continent to be here. In the campground parking lot were license plates from Missouri, British Columbia, Florida, Nebraska, Texas, and Vermont, as well as most of the far western states. A number of these vagrants were college students combining political action and a summer lark. Another contingent, weather-beaten and road savvy, was decorated with tattoos and ornaments of feather and bone, and carried carved wooden staffs or guitars. There were a handful of licensed witches and a group called the Redneck Women's Caucus. A sizable minority appeared to be earnest middle-class folk, only a touch rakish—Guatemalan prints, a ponytail or earring—but with the requisite shining eyes of believers. There were even two anthropologists with notebooks and safari hats.

I came as a freelance journalist. I had written a little in the past on logging and mining and Indians, and had lived for twenty years in northern California, so I thought I was well positioned to do the work. I had contacted the editor of a big New York magazine and received a guarantee of expense money for an on-spec piece, so I intended to play hard at being a professional eagle eye. I would scan the affair for intimate details, snoop for ironies and ambiguities, articulate the high-contrast images and resonant symbols.

There was plenty of material. Coming here on Highway 101 a motorist will pass hundreds of big rigs stacked with logs or lumber or fiberboard, and several large mills (Masonite and Louisiana Pacific in Ukiah, LP again in Willits, Eel River in Leggett, Pacific Lumber in Scotia). This relentless caravan of dead wood signals that one has entered the "redwood empire," and indeed, a few miles beyond Willits, one can detour a few hundred yards through the Avenue of Giants, a series of memorial stands named after wealthy citizens who purchased, and thus protected, a narrow corridor of gloomy, colossal old trees. Roadside attractions are plentiful at the fringe of these groves: tree trunks hollowed out to house boutiques and yards full of chain saw sculptures, including a gigantic golden bear (now extinct) rearing erect under a simple, straightforward sign: Carving for Christ.

Or take the little town of Scotia, wholly owned by Pacific Lumber, a subsidiary of Maxxam Corporation. Here are rows of neat company houses along clean tree-shaded streets, a company clinic and company gym, a tidy company market district, and all around in every direction more trees, four hundred square miles of them, mostly dense green second- and third-growth stands. And the company offices, plain but homey, where a resource manager, a calm articulate lawyer, mused over the mystery of a group of fanatics devoted to the raucous denunciation of this small-town American dream, and went on to explain to me that although Palco is "very sensitive" to the environment, it would actually be *illegal,* according to his reading of the statutes, *not* to harvest the last ancient forest in private hands. He also pointed out that the company cared deeply about habitat and wildlife, and voluntarily built its own hatchery to restore the winter steelhead run on the Eel River.

Going back to my car after this interview I passed along one border of the main mill, where three sixty-foot bandsaws sing and howl, cutting half a million board feet every day and night. Across the road I noticed a row of concrete tanks almost overgrown with bushes. They looked like roofless, half-buried handball courts, topped by a chain-link fence. Curious, I approached and found them completely empty. They had been so for a long time, a thick layer of dead leaves and debris scattered over their slab bottoms. A small sign was wired into the fence, identifying these dry tanks as, in fact, the steelhead hatchery of which the lawyer spoke.

Or to jump ahead a day or two, consider Judi Bari and the solar-powered golf cart. Rules were bent to permit this vehicle on footpaths of the ancient forest, because in 1990 a pipe bomb ripped out Bari's womb and fractured her pelvis, so she could not walk comfortably over uneven ground to the glades where her talks were scheduled. She was a hero and martyr to the environmental cause, and a tough, shrewd spokesperson for workers and women—a sort of hybrid of Dian Fosse and Emma Goldman. Her visit here would be greeted by the secular equivalent of the public adoration of a saint, which as it turned out was not inappropriate. (Though she was in fine and fiery form at this rendezvous, two years later cancer would kill her.)

Salty and sardonic as they come, Bari could very well have appreciated the irony of her transport in this ecologically correct vehicle designed for that water-wasting idle pastime of the leisured and wealthy, through a wilderness prepared for her progress by laying sheets of a laminated wood product across minor gullies and soft spots. We in the multitude, meanwhile, made no outcry over our plastic shit-buckets or the walkie-talkies used on security patrol. No one, including the most radical environmentalist or the deepest ecologist, avoids compromise with technology, and our mindscapes are rich in hinted oxymorons like solar golf carts and plastic waste containers.

Then, of course, there is the extensive literature generated by these inconsistencies, paradoxes, and ambiguities (recorded on the dead, pressed flesh of trees), for people are increasingly aware of and bothered by them. How we view nature, or should view it, what wilderness and wood products can do for or to us, whether we should kill and eat other creatures—these are questions fiercely debated now in universities, in fat hardcovers, and in dozens of scholarly journals. Third-graders in our public schools are learning to define "ecosystem" and "biosphere," and more and more of them will grow up to join the phalanxes of learned experts who contribute an annual tonnage of data and commentary to various draft and final documents assessing the environmental impacts of practically any human enterprise that creates a measurable disturbance. Finally, a generation of talented nature writers has jetted about the planet in search of experience and inspiration, in order to bring some of this intellectual ferment to the general public, and world leaders, including our presidents, regularly pay homage—how sincere is a matter of dispute—to "the environment" as a vital, a *global* political, social, and economic concern.

On the Trinity River I supplemented such intriguing ironies, striking images, and elegant texts with three days of seminars and workshops that were fascinating, maddening, laughable, deeply moving, pathetic, and weird: Animal Rights,

FBI Plots, Wilderness and Gender, Jesus as an Ecoactivist, Native Plants, Traditional Medieval Sauna, the Warrior Poets' Society, Guerilla Guitars, and so on. I drank and smoked with these itinerants, listened to their poetry, was suspected of being an FBI plant and confronted by tearful accusations of patriarchal tyranny, but was also hugged and serenaded. I recorded numerous interviews, took copious notes, raced through a couple of dozen books and articles, and wrote thousands of words, which the New York editor rejected with a swift and summary cruelty.

The editor wanted something short and sharp on "the scene." So I had tried, and failed, to condense and clarify an experience that would in the end, after several complicated turns, profoundly alter my view of this world. My agents of change were not the famous, fiery activists, or the brooding trees, or the adrenalin of confrontation, or any novel idea forged from the seminars and workshops. They were unofficial and entirely in the background. They were dogs, children, and cows. Dogs and children led me to Tony and the cows, and Tony, in the few months that remained of his life, led me to reconsider all my attitudes toward the natural world and our place in it.

The information packet mailed out before the rendezvous contained a tongue-in-cheek prohibition against "dogs with cameras," and it soon became clear that a long-running, acrimonious controversy was behind this joke. In the open discussion portion of the welcoming circle, lively and pointed exchanges took place between pet owners and the fiercest guardians of the ancient forest. One faction argued that dogs also needed to get back to their roots and bond to the wilderness, while the other retorted that a poodle was as repulsive a human artifact as a disposable diaper and did not belong in a natural cathedral. A majority seemed not to give a flying fig but enjoyed the dogfight and egged it on. The debate was good-natured, if edgy, and the smiling discussion leader was just making a move to round it off when another man stepped forward to speak.

I had noticed him from the first because of his size and flaming beard. He was six-feet-four or better, very wide at the shoulders, with the bearing of a Viking. In a steady and clear voice he said his name was Tony Merten and he wanted to make a public apology for his behavior at last year's rendezvous. He went on to recount briefly the particulars. He had always objected to the dogs and found himself assigned a camp spot next to a family with a large and unruly hound. He complained and the owner made a strong rejoinder; in a subsequent more vigorous exchange the dog owner's glasses were broken.

When he cooled off, Tony said, he spoke to the man and tried to make amends, and after the rendezvous sent a check to cover the cost of the glasses. He had never meant to create an embarrassing situation, and though he had not changed his mind about the inappropriateness of dogs in the wild, he realized his reaction had been too hot and had disturbed the harmony of the group. He understood that they needed to stand together in defense of the earth. He could only say that at that particular time he had been very upset and dejected about the level of environmental damage in the country, and the boisterous dog had been the last straw, so he hoped now that those who might have been offended by his action would forgive him.

The man delivered some of this account with a wry twist that brought smiles. But he was wearing shorts, and I happened to be near enough to see that his muscular brown calves were shuddering just perceptibly, like suspension cables in a gale. There was never a tremor in his voice, and he kept his hands behind his back in a kind of parade rest, but I guessed he was exerting a good deal of effort to bring himself through this admission and apology. A proud man driven to humility by his own principles has no easy job of it. There was a brief, dead silence after Tony's address, and then several people called out: "You got it, man. . . . No problem. . . . All *right.*"

The facilitator said, OK, that's it for today, but already the circle was breaking up, people aware that they had passed through a little climax of understanding and sympathy. It occurred to me that a sidebar on the dog issue, featuring this incident, might be useful—some pathos and comedy to offset the militant T-shirt celebrating the extermination of cowboys. But others had already gathered around Tony to say a few words of encouragement, so I drifted on, figuring to catch him later for a short interview.

Chance threw us together soon enough. EF! has no special slot for visiting journalists or scholars. Like everyone else they must register, pay the fee, and between workshops do a share of the camp duties. The power figures whom I had targeted for extensive interviewing had not shown up yet, so I volunteered to put in a shift at the children's area. I found myself paired with Tony in erecting a big tent. We had help from four or five older kids, who eventually figured out the bracing of the poles, and once adult muscle had gotten the structure up, they politely dismissed us. That left some extra time before the afternoon round of workshops, and Tony offered me a beer.

It was home brew, an excellent medium-bodied ale. We sat on the tailgate of his compact rig and drank two apiece while we talked. I found out Tony made his beer from scratch, using local ingredients, and ate mostly from his garden, which was in the New Mexico desert, a place he loved very much. I do too, as it happens. My wife is from a little town there near the border, and I mentioned that we visit her parents every year in Deming. That was just on the other side of the Florida Mountains from his place, Tony said. The Lunatics from Luna County. This connection pleased us, so we kept talking about ourselves.

Tony was a neat, self-reliant camper. It was no surprise to learn that he had built his own house and was currently adding a big greenhouse where he could give his vegetables an early start. He said he lived away from people by choice, and I gathered he had made enough money in the marketplace ("computers," he said tersely) to support himself in a simplified lifestyle. He could devote his time to what he cared about—the desert and the life that flourishes there. He was direct and intense, yet almost laconic. I heard nothing on the order of a sermon. Millennial seers were a dime a dozen at this congress, ranging all the way from the licensed witches to population control nerds, but Tony was the only reticent one I met. He told me in passing that he lived without hope, because he thought the momentum of industrial technology was irreversible and its direction was downward to apocalypse or chaos, but we went right on to a discussion of activist tactics and the relevance (or lack thereof) of Deep Ecology, and after that there were doomsday jokes and talk of maybe, come Christmas, trading my California almonds for some of his fine beer.

In July it was hot even on the Trinity and we made a move to the river. There were usually clothes strewn on the rocks and a dozen or more heads visible in the two deep pools near the encampment, but it was the lunch hour and we had the place to ourselves. We were dumb for a while, porpoising around in the blue sky and big fluffy clouds reflected on the cool water. When the conversation resumed it coasted free. We liked each other, maybe because my own fundamental skepticism was not so far from Tony's bleak vision, at least on the global level. I mentioned at some point that I admired his courage and openness in making that public apology.

He winced a smile. Of course he was embarrassed, he said, but that was nothing beside the despair that had driven him to break the man's glasses in the first place. He could live as lightly as possible on the earth, respect the desert around him profoundly, go to meetings and write letters and join demonstrations, but it seemed to make little difference. The great trees were falling to the saws; miners gouged out whole mountains; species went on vanishing; cattle devoured what was left of the range—but anyway we all knew these things, everybody did, we were repeating ourselves. The horror had become boring, which was in a way its final and most unbearable stage.

It was hard to have a life in such a world. He confessed he was lonely. The desert was beautiful and he would live nowhere else, but he wished he had a companion, a good woman. He had already noted how many attractive young females were showing up in these pools in the heat of the day. He couldn't help but hope for something. He hadn't been hooked up with anyone for more than a year, and his last potential romance had fizzled out months ago. He mentioned the name of one of the veterans here from the Southwest, whom I had interviewed the day before. He thought he was, at forty-three, too old for most of this crowd, and maybe his size put women off.

I told him he wasn't too old and he wasn't too big. He just needed time. It was admittedly hard to find someone free and interested, with ideas and a temper that fit, and on top of that willing to live at the end of a long dirt road. I thought—but never said—that it was possibly also true that Tony's intensity, the sheer force of his convictions, would narrow the field for him. In my experience, passionate idealists have a hard time finding mates; they are attracted either to doormats or other firebrands, and the former is now very rare and the latter an explosive combination. It was too bad the situation with the lady from Arizona had not worked out. She was one of the leaders in this company, a big handsome woman with a mind of her own. Or perhaps a woman with a mind of her own is always a problem for men, as some feminists assert.

CHAPTER TWO

Various techniques have been developed to render the teaser bull sterile or incapable of coitus, including penile translocation. Translocation at the proper angle makes penetration and effective discharge of semen virtually impossible, but reproductive and eliminative functions remain unimpaired, and the teaser bull is able to mount and mark cows without difficulty.
—Summarized from textbooks on large animal surgery

Willie tells us that cowboys ain't easy to love and they're harder to hold. Still, shooting them seems an intemperate rejoinder. In the aftermath of my Trinity River pilgrimage I was plagued by the memory of that lady's T-shirt, and my dismay finally worked itself into a mild outrage. Not that I had any naive, romantic notions of white knights in wooly chaps, brave and free on the lone prairie. On the other hand, like many of my generation born in the West, I counted cattle and buckaroos as part of my heritage, and the idea that some people openly yearned to wipe out both species was more than a little shocking.

My own grandfather punched cows on the Chisholm Trail and followed the trade north to Wyoming and then Idaho, where he homesteaded a little quarter-section cattle ranch and tried to support a wife and seven children. I spent some time on a horse as a youth, and eventually hung around a ranch in Montana long enough to learn that the work was not always what you would call idyllic. Eventually I married into another ranching family, and my father-in-law, Babe Penn, became a close and dear friend.

Babe was a working cowboy and occasional rodeo hand in his youth, a ranch foreman and horse trainer at middle age, and a dairyman and brand inspector in his last years. He departed from this pattern only once, in 1942. He joined the marines and fought across the Pacific until, jockeying an LST onto the beach at Iwo Jima, he took substantial fragments of shrapnel in the backside and had to come home on a hospital ship. But he was soon in the saddle again and stayed there until the very end of his life, when his heart was too weak to allow him to ride, even as picket on easy drives with his old rancher friends.

Babe was never cut out for retirement, and long after his bypass operation he continued to work part-time as a brand inspector. Two or three times a week he drove his old Jeep over the border to Palomas, the little Mexican town where he checked livestock coming in for export to the north. Hot and dusty in the summertime, the job was nevertheless easy, as outdoor work goes, and kept Babe abreast of news and gossip in the cattle trade. It also left him with time to indulge his other passion, which was reading. There is a popular notion that working men in the ruder trades are creatures of few words, who, if they read at all, consume mostly girlie and gun magazines. Babe was living proof of the incorrectness of this fancy. He digested all kinds of material, forged his way through whatever anyone pressed on him, and educated himself informally but very well. A short list of some of the books I sent him over the last few years would include *All the Pretty Horses, Cadillac Desert, Den of Thieves, With the Contras,* and *Son of the Morning Star.* He read fast and close and dearly loved to discuss and argue the merits or shortcomings of a new book.

Our discussions were often lively and extended. Babe was a formidable adversary, mostly because he confronted one with an inexorable tolerance and open-mindedness. He was so anxious to get into the other fellow's moccasins that, before you knew it, you were trudging barefoot beside him while he thoughtfully pursued this path that you had so confidently chosen, but that now, under his sharp gaze, began to appear narrow, uneven, and choked with stones and thorns. Finally he would halt, sigh, hand back the footgear and—apologetically, mournfully even—raise the question of just where the hell we *were* on this thing. At that point, more often than not, I no longer had the faintest idea myself.

Still we saw eye-to-eye on a number of major issues: we would esteem at similar and mirthful measure such cultural icons as Madonna and Microsoft, and held the Department of Interior at about the same level of contempt. Environmental questions also interested us both, and even here we were usually in agreement, but we stumbled over one topic that divided us more seriously and required a measure of wariness and delicacy to pursue. This was the simmering controversy over grazing permits in the national forests and on huge tracts of high desert in the Great Basin, supervised by the Bureau of Land Management (BLM).

In our preliminary skirmishes over this terrain I sensed a deep unease in Babe, a complex reaction of offended pride, shock, and bewilderment. He was, I think, reflecting accurately enough the milder sentiments of most stockmen in his area, and they in turn were representative of a current running throughout the West. Ranchers were nervous because the Clinton administration had appointed a notorious liberal environmentalist as secretary of the interior, and Bruce Babbitt had, as one of his first official acts, commanded the BLM and Forest Service to begin preliminary studies for the reformation of policies and regulations governing grazing on federal lands. In this context, early in the 1990s, some activist ecogroups were already urging much more than mere radical reform. They wanted to terminate grazing permits on public land altogether. *No Moo in '92* and *Cow Free in '93* bellowed the bumper stickers.

In a larger format, these partisans set forth harsh arguments in support of such a sweeping ban. They asserted that, despite a reputed detestation of federal regulation, a few thousand western ranchers have enjoyed a lucrative subsidy at taxpayer expense for generations, paying nominal, even derisory, fees to skim a profit from 300 million acres of public rangeland and leaving to the government the cost of restoring vast, depleted tracts of that land. Further, the critics charged that pursuant to provisions of the Public Range Improvements Act of 1978, a compliant Forest Service and BLM have regularly covered the expense of developing fencing, access roads, and water supplies for stockmen, and in order to provide public input to guide their policy decisions, these agencies have established advisory councils that are composed overwhelmingly of local ranchers. In exchange, the people receive only the bill and the dubious satisfaction of providing welfare for the western cowboy—a romantic legend of independence and self-reliance with little remaining connection to reality.

This position was articulated in sharp detail and in a fiery polemic style by Denzel and Nancy Ferguson's *Sacred Cows at the Public Trough* (1983) and gained powerful support in the Southwest when Edward Abbey, the original prophet of monkey-wrenching, delivered a scathing attack on the intrusion of livestock into wilderness lands. Various grassroots activists and local chapters of more conservative environmental organizations took up the cause, and soon court challenges and press releases were creating a stir in the media, bringing to a startled public the first hint of serious challenge to a primary national myth.

Twenty years earlier, when Ben Cartwright was entertaining all America at Ponderosa Ranch and Babe ran an operation on thirty thousand acres in Colorado, a critique of this sort would have been laughed out of court and off the pages of responsible newspapers. The cowboy had already been a revered American icon for more than a century, and the high plains were assumed to be his natural home. Summing up the attitude of the previous generation, Will Barnes observed in 1913 that "from the first settlement of this country the pioneers have used the vacant lands about them for grazing their stock with little or no supervision or restraint, and no scheme which presumes to lock up these grazing areas against the coming of the settlers' herds will ever meet with the approval of the American people." A half-century later, a retired federal land officer wrote of the region, "It is, above all, the land of the range livestock rancher. No other use has been found for most of the land of the Intermountain West." Even in 1996 a group of high-level scientific advisers merely translated the same conclusion into a polite tautology: "Domestic livestock grazing is the most widespread and extensive use of western federal rangelands."[1]

These statements seem self-evident to anyone who has driven the blue highways of the Great Basin, crossing those vast, desolate reaches of sagebrush and rock and sky where a tin roof, a row of poplar, and a scatter of cows are often the only human mark on the landscape for many, many miles. Between the old railhead cities—Omaha, Kansas City, Santa Fe, Denver, Cheyenne, Salt Lake, Boise, Billings—lies a tremendous patchwork territory, an area three-fourths the size of Mexico, inhabited permanently only by some thirty thousand ranchers and their families (and 6 million cows and a couple of million sheep) and the small communities of workers and merchants who serve them. This cow country has remained fairly stable for six generations, even though the population of the whole region has ballooned tenfold and more, as most of these far-flung raw railroad boomtowns transformed themselves into glittering metropolises encircled by suburban colonies.

During more than half that period western lands were open for the taking, and not only stockmen but also homesteaders, lumbermen, prospectors, and railroad entrepreneurs rushed to grab what they could. Most of the West, however, is *not* laced with gold or covered with mighty trees or irrigated by cool, clear water; it is a rugged desert with a climate of extremes, and nearly half a billion acres of it remains uninhabited public land. Franklin D. Roosevelt set aside the last significant parcel, "all vacant, unappropriated and unreserved lands" in the western region, by signing the Taylor Grazing Act in 1934. Even now almost 90 percent of Alaska, 80 percent of Nevada, nearly two-thirds of Idaho and Utah, and half of Oregon and Wyoming belong to the federal government.

It might appear that no one except the hardy stockman *wants* to live in these remote plains and mountain valleys, so that whatever the formal title and status of the land, it belongs to the herders who accept its hardships, survive there, and call the place home. The generous permit system (and nominal fee structure) established in the national forests by Pinchot in 1905 and on rangeland by the Taylor Act is thus simply an acknowledgment of reality. Babe thought that way, and it is, I would guess, the view most western ranchers hold of themselves and their region. When they mention a size for their place or operation, they commonly include permitted federal land, which more often than not amounts to several times the acreage of their deeded property.

The demographic profile of the area has been changing rapidly, however, and by the 1990s cattlemen, miners, and timber cutters could no longer shrug off the environmental movement as an insignificant handful of dyspeptic backpackers and bird-watchers. Many businesses have found the tax structures and prices of the West attractive, and they bring with them teams of lively and venturesome managers and workers eager to experience the "wild" West. Meanwhile a steady stream of refugee urban professionals and retirees arrives in search of spectacular scenery and a relaxed small-town lifestyle. Thus the fastest-growing states in the Union are Nevada, Arizona, Utah, Washington, Idaho, and Colorado. And of course tourism, both local and international, has increased markedly throughout the West during the last two decades.

Consequently what was once backcountry has been invaded by mountain bikers, rafters and kayakers, climbers, cross-country horse people, hang gliders, bow and black-powder hunters, rockhounds, nature photographers, mushroom and herb gatherers, and—last but certainly not least—the 30 million owners of sport utility vehicles who cruise forth to live out, in all-wheel drive, the advertised dream of mastering a mesa, flattening a mountain, blowing through a river in that newest and grandest of American theme parks—the wilderness.

When in 1994 the BLM finally released a document outlining its proposed changes, the ranchers knew their worries were not idle paranoia. The Draft Environmental Impact Statement on Rangeland Reform listed five alternative plans. Only one represented an improvement from the stock industry's point of view. Another held the status quo; two increased both fees and restrictive policies; and the last provided for the termination of all leasing on BLM lands. The ensuing uproar drove congressmen from the western states to begin hearings on legislation to short-circuit the bureau's plan for reform, and ranching communities across the West began to organize, lobby, collect war chests, and otherwise prepare themselves for a serious struggle, a defense, as they saw it, of both their rights and their way of life against a mighty coalition of slick greens, hostile government bureaucrats, and a public duped by a biased, liberal media.

On the far fringe of this reaction lurked the hobgoblin of armed resistance, a connection to the militia, home rule, and survivalist movements. In Nevada in 1994 a band of more than forty men, many with rifles, faced down a Forest Service ranger and a sheriff's deputy while a county commissioner bulldozed an access road across government land. Two years later the so-called Freemen, refusing to recognize federal authority of any kind, stockpiled guns, ammunition, and food on a remote Montana ranch and kept the FBI at bay for almost three months, threatening a fight to the death to preserve their autonomy (and avoid prosecution for tax fraud). Most ranchers do not advocate such extreme measures but might very well express sympathy for those who do. They are outraged at the proliferation of regulations and guidelines they view as either meaningless or harmful, and the insensitivity and stupidity of distant agencies: the Forest Service, BLM, EPA, FDA, and numerous state offices of water and air quality, fish and game, mining, parks, and so on. *Range,* a sassy, slick magazine out of Nevada (subtitle: *The Cowboy Spirit on America's Outback*), features personal accounts and opinion pieces from all over the West that detail this David-and-Goliath struggle. Some typical entries: "Fighting for Freedom," "Something's Wrong In Libby, Montana," "Rough Times in Diamond Valley," "The Great Grazing Battle," and "When Media Goes Green, Rednecks Get Guns."

The resentment in this community is doubtless fanned hotter by stories and rumors of ecoterrorism practiced or encouraged by groups like Earth First! and Greenpeace. Wise Use, a loose coalition against environmental organizations, regularly publishes reports of tree-spiking, vandalism, and arson directed at the resource industries, and in recent years has listed cases of the shooting of cattle on the open range. Two months before the Trinity rendezvous, Babe had told me of such an incident not far from Deming. Twenty-two head were methodically shot down in a remote corner of public land, apparently by a single sniper. This calculated slaughter upset local ranchers, for it went well beyond the occasional random act of vandalism, the overturned stock tank or cut fence.

Temperatures run high in land and resource disputes in the West, and always have. In the bitter range wars of the last century, men were routinely shot and hung, both legally and illegally, in disputes over stock, water, and land. Thirty died in the Tonto Basin War in Colorado, and half a dozen, including one woman, met lurid ends in a matter of days in Johnson County in Wyoming. Even today a large majority of the rural population of the West own guns and know how to use them, and it is something of a miracle that our ongoing conflicts over rangeland, timber harvests, and mining have not so far produced notable shoot-outs.

In September I sent Tony a couple of pounds of almonds with a short letter enclosed. I mentioned that in the spring we would probably be visiting in the Southwest and would give him a call. He answered a month later with a two-pager, talking about his work on the greenhouse, the perfect, exhilarating weather, and new projects he had undertaken, all going well so far. In a brief aside he noted that he still had not encountered a possible partner and was resigned to solitude for the time being. And of course he would be pleased to get together for a beer and some talk whenever we came to Deming.

As it happened, we abandoned our plan to visit the New Mexico relatives in the spring of 1996 and as a consequence were running up a good-size phone bill. Babe and I could be garrulous, so we generally talked after the women and children, pursuing our long-running arguments and speculations, which included the controversy over grazing on public lands. Sometime after Valentine's Day he opened our parley by announcing that the cattle-killers had struck again on another of the big ranches in the area, not far from the site of the first attack. About a dozen head shot. It was assumed the same parties were responsible, and this time there was a suspect.

Babe knew something about the investigation because his boss, Tom Bill Black, was running it. I was interested and asked for details, but there weren't many. The two ranchers who had so far suffered losses ran big outfits, several hundred head, and their operations would survive, but they were riled and a little spooked. Neither was involved in any serious disputes with neighbors or had any reason to expect a vengeful attack. Both were speculating that the shootings had some connection to local environmental campaigns, which had gotten more intense over the last year or two. Tom Kelly's Tres Lomitas ranch held grazing permits for thirty sections of BLM land, and Kelly was also the vocal president of a regional chapter of People for the West, an advocacy group for ranchers.

Three weeks passed before Babe and I talked again, and when I asked what had happened with the cow-shootings, he said he had some news. The case wasn't closed, but it might never be solved. The leading suspect had killed himself. Or so it appeared anyway. Found him at home in a chair with the pistol beside him and a bullet through his head. That was news all right, I said, and asked for particulars. That's about all of it, Babe said. A couple more things. His voice became subdued, and he went on with a reluctant, mournful tact. This guy lived way out, not far from where the cows were shot, and he was supposed to be a former president of the local Sierra Club. I thought that over for a minute and made some ambiguous, deprecatory sound. Privately I entertained serious doubt that any responsible member of such a mainstream group would be likely to take so great a risk. Any connection to covert action and gunplay would be a tremendous embarrassment. So what besides his location and his opinions made him a suspect?

He hadn't had much time to talk to Tom Bill, Babe said. The inspector was still wrapped up in the whole thing, but apparently he had gone to interview this fella about ten days before, and the man had acted pretty shook up. Tom Bill was doing some checking on him when he heard from the sheriff, who had picked up a tip from somewhere on a possible suicide. They went out there and sure enough, the guy had shot himself. It was a hard thing.

We talked on for a while, wondering. What puzzled Babe was why a fella would take such a step. The total of thirty or so cows—some actually calves—were only worth around fifteen thousand. In New Mexico that wasn't a felony offense. The man would have to pay it off, with a fine, and at the worst maybe do ninety days. I agreed the real motive might lie deeper—assuming the Sierra Clubber actually was guilty— but I was more curious about what the implications of the case might be for the environmental movement. Babe said he would keep his ears and eyes open; he knew I was interested and so was he, so if anything new turned up he'd let me know in one of our weekly calls.

It was less than a week before I received mail from Colorado from someone I did not know. Inside were two pages. The first informed me very briefly that Tony Merten had taken his own life and had left a final letter, with instructions to mail copies to his friends. From the first sentence of the cover note I guessed, of course, but an instant later I underwent a second shock of amazement, actually a double amazement: first, that I had never once thought of Tony, even idly and incidentally, in relation to the cattle-killings, and second, that now I would have to do exactly that.

His last message was brief and direct, but gave no answers, left me in fact with an overwhelming question. He wrote that despite being perfectly satisfied with his life—everything "going well"—he was "fearful and terrified of the near future," because the planetary ecosystem was collapsing. "I see no hope for humanity or the Earth," he wrote. "It is better to check out now than sometime later." Except for this passage, delivered flatly and without elaboration, there was no sign of stress or strong emotion. He asked simply that people in the environmental community and family members be notified, and then closed: "Tell everyone I loved them all. Thanks." His tone throughout was one of blunt reserve, which fit the character I remembered. I also found traces of familiar wry humor. After giving the date—"February 17, I think. (Whatever, it's Saturday)"—he opened with the line: "This letter is going to bum you out."

It did more than that. I went over and over my memory of that flame-bearded Viking, radiating health and vigor, honest in his speech, a tidy camper and fine brewer, self-reliant and proud, a man of intelligence and spirit. I could not square those impressions with the fact of his suicide, or the ugly circumstances surrounding it.

In the first days after this news I verified that Tony had indeed been chairman of the local Rio Grande chapter of the Sierra Club and was a respected member of other environmental groups as well. He had published letters and opinion pieces in local journals, and a wilderness alliance in southern Utah had already scheduled a memorial ceremony for him in September. He lived alone, but not as a recluse; he belonged to a community, one with an agenda and public profile.

I eventually spoke on the phone to a few of his friends. They were glum and circumspect, uncomfortable with what had happened—a mirror, I realized, of my own state. No one dismissed outright the possibility that Tony could have taken out the cows, but no one ventured to confirm its likelihood either. Among local ranchers, of course, opinion was very different. Babe kept me up to date on the views from this quarter. Although the investigation had turned up no hard evidence, and no formal charges or allegations were made, the fact that Tony probably shot himself the day after Tom Bill questioned him was held to be a convincing circumstance. In common opinion, the guilty party had identified himself, so unless there were more attacks or an unexpected discovery, the case was effectively, if not officially, closed.

CHAPTER THREE

In a sense, violence is a test of the sacred, a matter of what will or will not be defended when push comes to shove.

—Jack Turner, *The Abstract Wild*

The tautology gives us no way out: if wild nature is the only thing worth saving, and if our mere presence destroys it, then the sole solution to our own unnaturalness, the only way to protect sacred wilderness from profane humanity, would seem to be suicide.

—William Cronon, *Uncommon Ground*

It was a year before I got back to New Mexico, ostensibly to help Babe with a sewer line and a couple of leaky roofs on outbuildings, though we both knew the real reason was that his big heart was giving out and we wanted to see each other again. He was too ill to do any brand inspecting and so was there to lay out plans for me, hand me the shovel and hammer, and make coffee. We talked around the job, and in the evenings after the news, took up our usual range of topics. Of course we got back into the mystery of the slaughtered cows, and it seemed to me Babe remained almost as bothered and curious as I was.

He knew his own time was almost up, a life spent working with cattle and horses, and I think it troubled him in several ways that a healthy, strong man might have killed the cows and then himself out of despair provoked somehow by the ranchers' trade. With his usual tolerance he accepted my impression of Tony, that he didn't appear to be deranged or distraught. I think Babe viewed this opaque death as another piece of a whole sinister, incomprehensible pattern of events in modern life—youths shooting idly at passing vehicles, the CIA supporting bloodthirsty tyrants, million-dollar lawsuits over tasteless flirting. Such events meant the common sense he had always believed in could no longer be taken for granted, that something had maybe gone awry in the land he loved, had fought for, and intended to be buried in.

For my part, I had to admit I was still irritated, upset, and fascinated by this tragedy, even though it seemed certain the truth of the situation would never be known. Details were too few, the chain of assumptions too long and too frail. But now here I was, half an hour's drive—Babe's estimate—from the place where everything had happened. The next day we drove by Tony's house, a trim white bungalow with a high clerestory on the sunrise wall. We were on a gravel road, the nearest dwellings barely visible, miles away. The construction looked simple, sturdy, and unobtrusive, but except for the big greenhouse in the back, where Tony had been found, nothing indicated a common rural life: no rusty machinery, piles of scrap, or drums of petrol, no barn or chicken house. The place had been sold, Babe thought, but there were no signs of recent activity.

To get there we drove through a notch in the Floridas, one of several mountainous spines that rise suddenly as saw blades from this desert and run twenty or more miles before ending just as precipitously. A thin scrub of yucca and cholla and prickly pear crawled up the flanks, with a few juniper along the steeper crevices. Fences were rare, and I saw no more than a tiny, scattered herd of cattle on a distant slope. When I asked about Tony's place Babe said it was very small, a couple of acres. In recent years some ranchers had sold these little dabs to people who wanted to get away to the deep country. Not far away—he pointed southeast over the fifty or so miles of rolling desert laid out between us and El Paso and Chihuahua—one developer bought a big tract, several sections, with the notion of carving it into ranchettes, but the project never went anywhere.

This desert was so empty and so vast someone could readily find a lonely jeep track from which he or she could walk a few hundred yards up a slope or down an arroyo and shoot a small herd of cows. If one could locate them. Ranches in these parts might include up to eighty square miles of rangeland under permit. The agitation about overgrazing seemed slightly unreal in this context: a few distant moving dots inhabiting hundreds of thousands of acres of thorns and stone. Not even fences or water towers. Well, Babe explained, the fencing was part of the problem, one reason why the newcomers here got so upset with the ranchers. The way it worked, if you bought a place next to government land with a grazing permit on it, the rancher didn't have to fence in his stock; you had to pay for fencing them out. And you would be surprised, he went on, at how cows can get around to find forage in this open country.

That made one thing clear. The campaign to evict cattle from public land could have a personal urgency for those who purchase small properties near the desolate territories administered by the BLM. These newcomers might reasonably claim that as taxpayers they are already part owners of their surroundings. And they could well view cattlemen and the government as collaborators in an unfair exploitation of a common birthright—the relic of a time when there was more open space than most citizens could stand, and only herdsmen, for practical reasons, cared a whit for the lone prairies. Tony had told me how he loved this desert, and it must have jarred him to see from his doorstep the cows probing his space, as well as territory some financier coveted for subdivision and quickie construction.

It seemed a natural next step to talk to Tom Bill. Babe kindly called him and explained my interest, asked if we could meet for an afternoon session. It was not a problem. Things were a little slow, not much coming over the border, so the inspector could drop by tomorrow for an hour or two before dinnertime. He was, Babe said, looking forward to it. The next day was warm for late March. After shaking my hand Tom Bill regretfully chose coffee over a beer, remarking that it was pretty plain he had enjoyed, as a youngster, a close enough acquaintance with brew. He tipped off his hat then and eased his considerable presence into a chair at the kitchen table, where Babe and I were already set up. He asked Babe how he was doing, and the two of them chatted for a while about office matters, the cattle trade, and so on. Tom Bill was not one of your laconic, steely cowboys. He took his time, but he was a genial man who liked to talk and was good at it. When I reached for my notebook, he was there and ready. We talked through that afternoon and part of another, and finished with a couple of long phone calls later.

In his mid-fifties, Tom Bill was born in Silver City and worked as a cowhand before becoming a brand inspector twenty years ago. It was, he mentioned, the same career his father had followed. He had been at the BI's office here in Luna County since 1978 and knew the territory well enough. He liked the job, though it was not always as challenging as the case we were here to discuss, and as supervisor he spent too much time indoors or in a vehicle. He explained that brand inspectors were, in New Mexico, fully certified law enforcement agents, which meant he had primary jurisdiction in the investigation of these cattle-killings. And this was a highly unusual case from the beginning.

The first incident, two years ago on Tom Kelly's place, had shocked him a little. There were no apparent vehicle tracks, and the herd had been scattered out pretty well. The killer had worked in a circular pattern, picking off twenty-two head, including some cows with newborn calves. It would have taken at least an hour or two of methodical stalking. The shooter left some cartridge cases, but that was no particular help because they came from a common weapon: an

SKS 7.62 x .39, a cheap Chinese copy of a Russian assault rifle. The carcasses were fresh, so Tom Bill figured the shootings happened right around Easter Sunday, which might or might not mean something. It was reasonably clear the motive wasn't a practical one, in the ordinary sense. Kelly, a retired army officer, also reported a drained stock tank and damaged windmill in the same area, and told the *Deming Headlight* he suspected the shootings and sabotage were the work of "environmental zealots."

Although he made inquiries among nearby ranchers, hoping someone had spotted an unfamiliar vehicle on the back roads, Tom Bill had no particular reason to believe the shooter might be in the immediate neighborhood. Deming and Las Cruces or even El Paso were all an hour's drive or less, and once in a while bikers, bird-watchers, or backpackers from those population centers trailed through. He had ruled nothing out but didn't possess the staff to systematically interview every resident in the area. And though he drove by it more than once, he did not stop at Tony's place, just four miles over the hill from Kelly's.

After ten months the case seemed no nearer a solution. No phantom group or anonymous guerrilla had taken credit for the action. No informant stepped forward to claim the six-thousand-dollar reward offered by the ranching community (one contribution coming from the TV commentator Sam Donaldson, who ran cattle in the area). Only one credible suspect had materialized, a man who had been seen shooting in the area a short time before the killings. Although he had an impressive arsenal in his home, including some SKS rifles, none of the weaponry checked out in lab tests. The man was only an avid gun fancier who had been target-shooting in the wrong place at the wrong time.

Tom Bill had explored the possibility that the shootings were an act of "ecotearism" (as he pronounces it when animated). He talked to the FBI, to a sheriff's office in Arizona with files on monkey-wrenching, and to Barry Clausen, whose book *Walking the Edge* purports to document the dire threat of Earth First!-style activism in America. He found this material interesting, a new set of profiles and signatures to consider. A brand inspector does not ordinarily attack crime from an ideological angle, but this was, after all, the era of the Unabomber, and the things Tom Bill was hearing from these sources made sense to him (though he thought Clausen exaggerated his case). Clearly there were people who strove fanatically to protect wilderness. They would, they said, put their bodies on the line to save old-growth forests, wild rivers, whales, or even deserts. Right here in southeastern New Mexico some environmental groups were vigorously challenging the ranchers' moral authority to graze public lands. Tom Bill found it significant that, according to the map, every one of those twenty-two cows had been picked off in a BLM section identified as a "wilderness study area."

Then, sometime around Valentine's Day 1996, a dozen more cows were gunned down on Bill Smyer's ranch, only a few miles from the site of the first incident. When Tom Bill and his supervisor, Henry Torres, arrived at the scene in the late morning of February 15, they soon turned up evidence that the same party or parties was behind this second round of killings. More shell casings from an SKS, and the same careful, straightforward execution. This time, however, the carcasses were more scattered and in different stages of decomposition, indicating that the shooter had made several assaults over a week or so. In an area of sandy soil near one kill they encountered their first significant break: a few footprints. The first thing Tom Bill remarked was their size, fourteen inches long on his tape. Also the sole showed traces of a pattern, indicating a sneaker or hiking shoe. The soil was too fragile for casts, so they took pictures. The layout of the tracks and their scarcity indicated that their owner had made an effort to keep to hard ground, so as to avoid detection. Tom Bill managed to follow the trail for most of a mile, until it approached a road. From the point at which they lost the sign, the two inspectors could see three or four houses. They set out then to talk to these neighbors, hoping someone had heard shots or seen a strange vehicle in the last few days.

The third house they came to was Tony's. He met them in the driveway, wearing shorts and a tank top. Tom Bill was surprised to see somebody he recognized. A few months earlier he had been called to present himself for jury duty and had noticed the large man with red hair, also in the pool of potential jurors. They even sat near each other, Tom Bill behind Tony. As it happened, neither was chosen, so they had not crossed paths again. Tom Bill's impression was that Tony did not remember him. The inspectors introduced themselves and explained the reason for the call. A conversation took place there under the bright desert sun and lasted maybe half an hour. At the first opportunity, of course, Tom Bill looked down. He saw very big feet in heavy-duty sneakers. It was easy for me to imagine how affable and alert Tom Bill would be as he went on to ask the routine questions, his attention shifting only now and then as he mused again over the surface of the driveway, which unfortunately was hard gravel and took no impression.

His subject did not appear nervous or hostile, and Tom Bill was struck by his forthrightness. Tony said he had been at home but neither saw nor heard anything out of the ordinary. Tom Bill asked if he had, himself, any guns. No, Tony replied. Except for a 9-mm pistol, which he kept mostly because he had been robbed a couple of years before. In fact his rifle had been stolen then, along with a mountain bike. A Russian-type gun, an SKS. Had he reported the theft? No, Tony said. Nothing was ever recovered. As for the business about the cows, Tony went on—without being asked, Tom Bill emphasized—he was himself an environmentalist, a strong believer in maintaining native species. Tom Bill pursued this line of discussion, expressed his curiosity about such beliefs. They talked about population, certainly part of the problem. Tony held that a lot of men ought to get themselves sterilized. It was also crucial to relieve the massive human exploitation of nature: Tony thought wild lands ought to be reserved for wild creatures. Cattle, he remarked, were European in origin and did not belong on this continent.

Would he, Tom Bill asked off-handedly, consider shooting a trespassing cow? Tony answered by pointing out that he used his shovel all the time on another unwanted intruder, Russian thistle. Well, Tom Bill said, weeds were a different case. No, Tony insisted, a tumbleweed chopped was also a life taken. The point was that people had introduced damaging alien species into systems where they didn't belong. By the time the conversation wound down, Tom Bill had privately noted several words and phrases that chimed with the information he had gathered from the Arizona sheriffs and the FBI. He was hearing fragments of what might be "ecotearist" language, as these sources described it. He was now definitely, genuinely curious about this big, red-haired, outspoken man and the beliefs he was propounding.

Tom Bill's moustache has a downslope of pensive sadness, and his smile is offset a little by watchful blue eyes. He had reached the point in his narrative when there were obvious grounds for suspicion, and one might have expected the reminiscence to take on a certain anticipation or even relish. But Tom Bill avoided the word "suspect," speaking simply of "Tony Merten" or "the man," with a kind of rueful fatalism. I interrupted to ask how he *felt* about his subject after that first interview, apart from his professional assessment. Well, he said, in this business you try not to jump to conclusions one way or the other. Some details were of course beginning to line up. Those shoes looked like they might fit the tracks, and the man seemed to be in terrific shape, certainly able to walk or jog six miles across familiar country, even in the dark. Then there was his frankly expressed dislike of vagrant cows and the somewhat oblique response to the gun question. On the other hand, personally speaking, he could see Tony Merten was educated and articulate. The type of guy who kept to himself and thought a lot. Had his own ideas, definitely. Oh yeah, Tony Merten was a very interesting man.

The day after this first encounter another brand inspector and a trio of ranch hands from the Smyer place were sent out to expand the perimeter of the search for cow carcasses, while Tom Bill went hunting for more tracks. This time he was accompanied by Arnold Chavez, an investigator from the district attorney's office. By late morning they had found a few likely matching prints, the last only a few hundred yards from Tony's place. After reclaiming their vehicle, they headed back on the gravel road that ran right by Tony's. Tom Bill says he had no plan to attempt another interview so soon, but they saw Tony outside in his yard and Chavez proposed on the spur of the moment to stop and talk to him. They pulled in by the fence, and the ensuing conversation took place across it. From the outset it was not a relaxed or amiable exchange. Tom Bill noticed immediately that Tony wasn't wearing the sneakers anymore, and he appeared defensive in answering questions. Yes he walked this desert often, miles every day usually. Yes he knew the country well. He loved this country. The discussion returned to the environmental issues pursued the day before. Later Tom Bill would recall that his subject had observed in passing that humanity, in its current state, would destroy the world in another twenty years.

Yet Tony did not break off the conversation or ask his visitors to leave. He began to complain of harassment. "He said we were singling him out because of his views," Tom Bill said, "because he was an environmentalist." This was of course becoming partly true, and again Tom Bill was struck by how honestly—even recklessly—the man upheld an opinion. They kept the discussion moving, and Tom Bill believed it was headed somewhere. "We had him going our way. If we kept talking, we would learn more and more." But at that moment another car pulled in.

It was Bill Smyer, who had been cruising the area checking on his riders and had seen the investigators' vehicle. When the rancher identified himself, Tony said, "Oh, *you're* Bill Smyer," and came forward to shake hands. The two began to talk, and very swiftly "some disagreements" were evident. These were about pretty general issues—two people from different planets—and soon the discussion was "all over the place." Tom Bill and Chavez tried to bring the conversation back to the previous track, but the situation got even worse. The team on horseback, Smyer's hands and the other brand inspector, picked that moment to ride up to see what all the cars were doing at the white house. At this point, Tom Bill said, Tony was pretty shook up. Which was, he hastened to point out, understandable. Seven men, three of them in law enforcement, staring at you over your front gate while you fend off uncomfortable questions. From Tom Bill's perspective, the opportunity for real communication had evaporated. The man was spooked and feisty, and there would be no new revelations that day. In this tense atmosphere, the session collapsed and the company dispersed.

Tom Bill was uneasy at this outcome. They hadn't intended to apply brute pressure, and now the man would clam up and stay on guard—whether or not he was personally guilty of the crime. Hence they were being hurried into the next phase of the investigation—assembling enough facts to justify a search warrant. Without witnesses or physical evidence, that would be a considerable chore.

Tom Bill ran background checks. Tony Merten was indeed a dedicated environmentalist, a former officer of the Sierra Club, and a member of other green organizations, both regional and national. It looked like that was his whole life. But he had no record of arrests, and his Earth First! association did not include any grandstand challenges to the law. He was open and assertive in his views, and in the most recent publication of the Rio Grande chapter of the Sierra Club, he had written a dry and sardonic piece on his own role in the region. "I am a public lands rancher. Some hundred and sixty head graze several thousand acres near a spring on my land in the Chihuahuan Desert of southern New Mexico. They graze for free—I don't pay a dime to lease the land or its forage, even though I own less than one-hundredth of one percent of it. And there is plenty to go around—inside my fence, a dozen species of herbs and forbs grow here and nowhere else. That's because the herd aids in seed-dispersal and soil fertilization. And the springside riparian community has rebounded since I bought the place four years ago." This beneficent "herd" consisted of the local mice, rabbits, and deer, not "exotic, destructive species imported from Europe." Tony pointed out that his fences and spring, developed at his own expense, were providing for the public what the government failed to deliver—habitat for wildlife, a refuge from the ravages of cattle-grazing. He wrote of himself as the protector of this public trust. "And I do not own them. I care for them. I am a steward of the land."[2]

Tony's claim to have once owned an SKS rifle also turned out to be truthful. The purchase was registered in October 1993. He had also vouchsafed that he didn't report the theft of his rifle and mountain bike at the time of the burglary, in the spring of 1994. But Tom Bill remembered Tony had been mustered into the pool of jury candidates the following year, remembered also that those called must fill out a questionnaire. On that form candidates are asked if they have ever been the victims of crime, and if so, they must give an account of the occasion. Tom Bill went to the courthouse and pulled Tony's forms. The key question was answered affirmatively, but with a remarkable omission. His home had been burglarized, Tony explained, but he listed only a single significant loss. One mountain bike.

Meanwhile, the cartridge cases collected from the Kelly and Smyer ranches had been sent to the lab, but the results were inconclusive. The forensic expert would testify that all the samples came from the same caliber and type of weapon, an SKS, but he could not certify that both sets of casings had issued from a single rifle. There were not enough signature markings. Strictly speaking, then, the investigators had no concrete proof that the two incidents were related. Nevertheless Tom Bill took what he had to the DA and urged that a search warrant be drawn up. After some discussion back and forth, the request was denied, or rather suspended. Tom Bill admitted there were reasons. He had not been able to measure or see the soles of those sneakers. The failure to record a legally registered, allegedly stolen rifle on the jury questionnaire could have been an oversight, though an improbable one. The background stuff was at best only suggestive and demonstrated no criminal or antisocial behavior. No witnesses or potential informants had come forward. Still, Tom Bill felt they were missing their best opportunity.

Other factors may have played a part in the DA's decision to delay and dig for more conclusive evidence. The cow-killings were a major media event, not only in Luna County but also in the whole southern end of the state. Even before the killings environmentalists and ranchers were running hot over another New Mexico "range war." Those words made the blunt headline on a front-page story in the *Las Cruces Sun News* on Sunday, February 18, just one day before the story of the shootings on Smyer's place broke. Gila Watch, an environmental group, was putting legal pressure on the Forest Service to withdraw a grazing allotment allowing cattle to run on the Gila and Aldo Leopold Wilderness Areas. A century-old ranch, the Diamond Bar out of Silver City, held the permit, and its current owner warned the reporter that if the rangers came to shut down his operation, "they had better bring a gun."

In such a climate the Luna County prosecutors had obvi-ous reasons for proceeding with caution. If the investigation stalled, the ranchers would be frustrated and angry, true enough. A week after the tense scene at Tony's house, one local cowboy had given Tom Bill a veiled threat: "If you don't get him, I will." On the other hand, if investigators identified a suspect but the DA failed to generate an arrest and trial, the cattlemen would be even more furious, and the environmen-talists would have a potential martyr, or possibly grounds for another lawsuit. The officials in charge wanted to make very sure they nailed the right man, and nailed him good.

The whole debate was, to the chief suspect, already moot. Tony Merten was dead less than forty-eight hours after Tom Bill walked away from his gate for the second time. The brand inspector never perceived any shadow of this impending trag-edy, though he did try to pitch his case for a warrant one more time, with the same result. He had gathered every con-ceivably relevant document, called every possible source, pored over the file for days. He found nothing persuasive enough. Two years later, recalling that time, he mentioned his regret that no tiny, clinching detail had emerged at the outset. He might have saved the man from himself.

On the afternoon of February 28, twelve days after he had last seen Tony, Tom Bill got a call from the Smyer ranch. It was Wednesday, sale day, so riders had been out the day before and that morning, combing the brush for cattle. They had been within sight of Tony's place more than once and remarked no lights, no smoke, no movement. His vehicle stayed parked. Maybe the man had skipped out, or something was up. Tom Bill and another inspector drove over the Floridas and took the road past Tony's. The place looked deserted, all right. They had decided to check with neighbors when they heard a deputy sheriff on the radio, asking his dispatcher for help in locating the Merten place. Tom Bill cut in to notify them he was already in the area and to ask what the problem was. The deputy had been sent on a "welfare check"; someone from out of state had phoned to report a possible suicide.

Tom Bill requested a rendezvous well before the turnoff to Tony's road. Neither he nor the district attorney had mentioned to the sheriff's office that a suspect in the cattle-shootings had materialized, since the inquiry had not progressed far enough to warrant any sort of alert. But the coincidence of the report from Smyer's riders and the call for a welfare check made Tom Bill cautious. It was not impossible, he thought, that the man had barricaded himself in his house. He might be armed and desperate. Far-fetched perhaps, but he didn't want the deputy barging into any sort of trouble. After a roadside conference, a second sheriff's car was summoned and the officers proceeded to Tony's place. It was nearly dark and they came in from four sides, but nothing moved. As soon as they entered the yard the odor hit them, and within a few minutes they located Tony in the greenhouse. He was sitting on the floor, braced by a camper's backrest, the 9-mm pistol still in his hand. He had apparently died of a gunshot wound to the head. Beside him there were also two empty boxes of sleeping tablets.

Despite appearances, Tom Bill roped off the area and called the DA and sheriff's captain. He remembered the threat from the cowboy a few days before and thought it wise to take no chances. Until they had looked the place over carefully, it would be treated as the scene of a possible crime. In the house they found the suicide letter, dated the seventeenth, and in Tony's pickup a receipt for the sleeping tablets, purchased in Socorro on the same day. Tom Bill kept looking. He was still the chief investigator, and the special, bitter circumstances of this emergency had given the brand inspector what the legal apparatus could not: one fleeting chance to look for new clues in a mystery that had preoccupied him for nearly a year, to develop (or eliminate) the only lead in a most unusual case.

He tried porches, closets, and the bedroom, and found plenty of footgear but not the sneakers he had seen two weeks ago. No rifle either, but he had expected that. He spent some time in Tony's study, because there, right away, things stopped him and required a second glance. In the typewriter was a half-finished letter. It was an imaginary exchange between Tony's cat and President Clinton's cat on environmental issues. On the desk was another letter to a physician concerning a controversy over charges for a vasectomy, a copy of the invoice attached. Tom Bill remembered Tony's remark about this solution to the population problem and had one more bit of proof—certainly not needed—that the man acted on his beliefs. Various official documents and government reports occupied shelves, and those that caught Tom Bill's eye were

CHAPTER THREE

of course the ones dealing with grazing allotments. He rifled through the stack. As brand inspector, he knew every outfit in the area, and he found copies of all the current BLM grazing permits in southwestern New Mexico, with two exceptions. The allotments for Smyer and Kelly.

Tom Bill left everything in place. He had found nothing substantial, one way or another, yet the mystery had deepened. The pharmacy receipt indicated that Tony had driven the three hundred miles round-trip to Socorro, brought back sleeping pills, and shot himself a few hours later. The argument with the doctor and the letter to the Clinton cat seemed to indicate that an ongoing, active life had been suddenly, dreadfully interrupted. The omission of Smyer and Kelly from the records of grazing allotments was at once tantalizing and mystifying.

The inspector had thought about and fretted over this case for a long time. When I asked if his concern went beyond the usual measure, he smiled his rueful smile and said, "I was *consumed* by Tony Merten." Even there in Babe's kitchen, long after all realistic hope of resolution had receded, Tom Bill was concentrated and alert, caught up in the scenes as he analyzed them again. There was perhaps more at stake than professional pride, a curiosity beyond the simple question of guilt or innocence. In his own mind he was pretty sure he had been on the right track. Not so much because of the footprints, the activist background, and the equivocal tale of the stolen rifle, but because he judged Tony Merten to be the "kind of guy who had a real tough time lying." His manner

51

seemed to betray him, not through a furtive nervousness or botched fabrications but quite the reverse: in his frank revelations, his provocative hints. Tom Bill thought it very likely the man did the deed; he also thought there was a chance Tony could have been talked into owning up to it, maybe at some level *wanted* the world to know. That, it was clear in his telling, was the outcome Tom Bill would have much preferred.

When I asked him to speculate on motive, the inspector was reluctant to say. It was of course possible Tony had nothing to do with the shooting of the cows and killed himself for other reasons entirely. Or if he was guilty of one or both assaults, it didn't mean necessarily that the fear of discovery alone would drive him to take his own life. He was a smart guy; he would know the crime was not a felony. If he had to make a guess, Tom Bill finally said, it looked as if Tony was lonely. His whole life was the environmental stuff, going to meetings and conferences and so forth, and working on his place way out in the desert. Maybe he wanted to impress others in the environmental movement, do something big time. But then found out he couldn't handle it, wasn't cut out for sneaking around. Woke up, realized the consequences, the way a guy will after a drunk, feeling real bad and guilty, even when you haven't done anything too dangerous, because your life just seems headed the wrong way.

Tom Bill also said he thought he understood what was bothering the environmentalists. What they said about population was true enough. Hell, everyone could see that. Too many of us using up too much—trees, minerals, land, water, whatever. And there were abuses on some grazing allotments, a person could find that. But most ranchers understood that depleting the range would never pay in the long run and tried to run a sustainable operation. Some had been doing it for three or four generations. Mainly he couldn't agree with the tactics the radicals used to make their case. Introduce guns and interfere with people's livelihood—then you've got a real dangerous situation, and nobody will care what your point is any more.

One odd incident suggested Tony was not unaware of that hazardous option. Tom Bill remembered a tip that had come in after a follow-up news story hinted at a possible link between the suicide and the cattle-shootings. A guy called up with this tale about Tony sending him a T-shirt. Guy said he recognized the name, and he'd had a lively discussion with this Merten at some public meeting or other, and just a few days after that Merten sent him a T-shirt in the mail. There was a picture of some Indians on the shirt and a line about how they were great for killing cowboys.

The case of the Luna County cattle-shootings has remained open, at least nominally, and when I checked with Tom Bill in the spring of 1998 there was a flicker of a possible new lead. The previous winter, two more cows were shot on rangeland out of Deming, and the youths responsible were caught. Teenage pranksters Tom Bill called them. But during his interrogation one boy protested that others had done much worse, that some kid he had seen around had shot twenty cows a while back. Tom Bill had pursued this tale, had assembled the high school annuals for the last few years and urged the boy to supply some positive identification, but nothing had come of it. The inspector wasn't hopeful the case would ever be solved, but then you never knew.

CHAPTER FOUR

And God said, Let us make man in our image, after our likeness: and let them have dominion over the fish of the sea, and over the fowl of the air, and over the cattle, and over all the earth, and over every creeping thing that creepeth upon the earth.
—Genesis 1:26

Cattle are mentioned fifty-one times in the Book of Genesis, compared to only seventeen references to sheep and a lone goat. Most often they are distinguished from all other beasts of the earth and creeping things, because they serve as a recognized measure of wealth, one pillar in a trinity of affluence: "And Abram was very rich in cattle, in silver, and in gold" (Gen. 13:2). The Lord also utilizes cattle to express his special favor or wrath, as when he guarantees to Moses that herds of the children of Israel will prosper and those of Egypt shall perish (Exod. 9:4, 9:6), or when he reserves good pasture to the descendants of Reuben and Gad, provided they cross the Jordan to wage holy war (Num. 32:1-29).

This valuation is by no means unique to the Hebrew tradition. Earlier civilizations esteemed cattle even more highly. Narmer-Menes, the first great king of Egypt, established a cult of bull worship that endured more than a thousand years, and neighboring civilizations—Sumerian, Mesopotamian, Phoenician, Hittite, North African, Minoan, Greek, and Roman—absorbed and extended that tradition. Hence Jehovah himself had to contend with the false god Baal, the Golden Calf; and Christ in his turn vied with Dionysus the Bull-horned and Mithra—though he did so by aligning himself with the very same venerable mythology of pastoral sacrifice, as the Lamb of God.

Cattle as gods, as riches. The progression is preserved in the etymology of the word, which derives from Medieval Latin *capitale,* "property" or "capital," which itself descends from *caput,* "head" or "chief." In India the cow has maintained something of its sacred status, and in Africa some tribes still figure wealth and prestige in terms of cattle. The Bantu, for example, traditionally exchange cows for wives. The bovines from these regions, especially the Cape buffalo and two strains of wild cattle in India, can indeed compel a measure of respect from humans; they are large, agile, and authoritative in temperament, not so far removed from those originals that served as symbols of godlike power and virility.

The ancestor of modern beef cattle was the European wild ox, known as the aurochs *(Bos primigenius),* a fearsome specimen measuring up to seven feet high at the shoulder. The aurochs survived until the end of the first millennium A.D., but its genetic legacy was swiftly adulterated and altered. The descendants of the wild ox have been bred for placidity, speed of reproduction, weight-gain, and tolerance for incarceration.

The principles of modern animal husbandry, developed in the last three centuries, have thus produced, in the Americas and Australia, huge herds of meat machines far more efficient and adaptable than the cattle domesticated by Neolithic cultures over several thousand years on the continents of Asia and Africa. The Herefords, Angus, Charolais, and Simmental (and their various crossbred permutations) that range over North America do not invite association with divine mystery and bear little resemblance to the swift, wild behemoths that inspired ancient poets and seers. Only certain sport animals—the longhorn from Texas and the Brahma bulls used for rodeo stock—retain a shadow of that primal presence.

So-called fat cattle may have lost their glamour, but they remain a source of considerable wealth in the world economy. There are about 1.3 billion head on the planet, more than a hundred million of them in the United States, which imports another couple of million, mostly for hamburger. The U.S. food service industry (restaurants, fast-food chains) feeds a hundred million people daily and takes in revenues of more than $200 billion every year, accounting for 40 percent of all the meat consumed in the nation. The business also employs 8 million people, more than any other single industry in the United States.[3]

For some, this great enterprise is a shining example of the evolution of human society through advances in science and technology. Where once tribes had to move their semiwild herds seasonally, and families spent much of their energy tending, protecting, slaughtering, and cooking or preserving the

animals they owned and relied on for survival, modern consumers can drive up to a window and, in a few moments,
receive a quarter-pound of hot, fresh, spiced beef with cheese,
bread, and trimmings, delivered in a sanitary bag with condi-
ments and a napkin. For this service they will pay a mere
fraction of their day's pay, and they need never see, hear, or
smell a whole animal, alive or dead.

This inexpensive, universally available, high-protein food
is the pride of the modern cattle industry, the consequence of
a technology infinitely more productive and efficient than the
old ways. Predators have been systematically exterminated.
Veterinary procedures and vaccines enhance reproduction and
survival. New systems for harvesting and storing feed grains,
transporting and holding animals, and irrigating and rotating
pastures allow a very small number of employees to control
large herds, and ensure that cattle grow and fatten as quickly
as possible. In addition, industrial chemists have found inge-
nious uses for the former waste products of the slaughter-
house, far surpassing the resourcefulness of primitives who
only made clothing, shelter, cordage, adhesives, and bone and
horn tools from their kills. John McPhee has compiled a cata-
log (not exhaustive) of this miraculous transubstantiation. "In
their soft, tanned appearance you can see the belts and brief-
cases. There is chewing gum in a cow, soft cartilage for plastic surgery, floor waxes, glues, piano keys. There are deter-
gents, deodorants, crayons, paint, shaving cream, shoe cream,
pocket combs, textiles, antifreeze, film, blood plasma, bone
marrow, insulin, wallpaper, linoleum, cellophane, and

sheetrock."[4] One may not think of worshiping a cornucopia of bubble gum and antifreeze, but it is surely fair to say that cattle have been integrated into the very fabric of contemporary human society. And the reverse is equally true: humans manage every aspect of bovine life. Cattle may lead a more artificial existence than any other species, with the possible exception of chickens.

For example, their reproduction is commonly manipulated at every stage. Teaser bulls, their penises replumbed sidewise to assure impotence and their undersides coated with a bright orange chalk, can detect and mark cows coming into heat. Or a whole herd of cows can be put on a synchronous cycle of estrus by receiving ear implants and injections, so that large numbers of embryo donors and recipients can be processed at once. Donors are given powerful hormone injections that induce superovulation, meaning the ovaries produce as many as twenty eggs in one cycle, instead of the usual one. Before recovery the eggs may be fertilized from frozen semen, collected by tricking a bull into ejaculating into a plastic sock, sometimes positioned in a motorized, hide-draped robot. Both egg and semen will have been selected from animals whose progeny show exceptional merit as milk or meat producers, and since breeders normally succeed in transplanting half the embryos obtained to recipient surrogate cows, production of the most desirable offspring is quintupled, in a conservative estimate, at the outset. A prime donor cow, however, can undergo superovulation every sixty days, and is therefore capable of supplying anywhere from thirty to fifty new calves a year.

After calves are born, this selection and modification process continues relentlessly. Some of the heifers will be tagged as breeders, egg donors, or recipients; the rest will quickly be processed into hamburger and sheetrock. Of the males, a rare few—the biggest and most vigorous—may become famous sires. They will receive dashing names like Renegade Victor, Red-eyed Jack, or Round Oak Rag Apple Elevation, and their influence, as jackstraws of sperm frozen in liquid air, will reach to the far corners of the earth. A few other bulls will breed small herds straightforwardly or be bent into teasers.

The bulk of the calf crop will be castrated, grazed for a season, and trucked to a huge yard where their only duty is to stand in their own excrement and eat. (Their food will be a carefully blended mix of grains, vitamins, antibiotics, growth hormones, syrup, recovered and processed agricultural wastes, and roughage, sometimes including a measure of plastic pellets.) At about eighteen months, an age that corresponds to early adolescence in humans, they are butchered, graded, ground, "restructured," injected, dried, dyed, rendered into broth or fat, preserved, canned, shrink-wrapped, frozen, irradiated, and sold. In one dish or another, Americans will eat a million head of these dismembered creatures every four days.

This scenario of efficient generation of foodstuff on a gigantic scale does not inspire or gratify everyone. Some environmentalists find it a nightmare vision, a lurid example of human mismanagement and ecological destruction. They charge that cattle and sheep have devastated the lands, waterways, and wildlife populations of every inhabited continent, annihilating as well the hunter-gatherer cultures that depended on untrammeled ecosystems. They argue that overgrazing is largely responsible for the accelerating desertification of the planet, that methane emitted from feedlots contributes significantly to global warming, that biodiversity is diminished wherever the "hoofed locusts" range.

The livestock industry, by this analysis, is an ecological catastrophe, yet for some critics the tragedy does not end there. They point to a final, obscene irony. The justification for raising beef must be its value as nourishment, how fully and efficiently it satisfies hunger. Yet, say writers like Lester Brown, Paul and Anne Erlich, and Frances Moore Lappé, the worldwide trend toward converting cropland to the production of fodder for cattle only intensifies famine in the poorer nations of Africa, South America, and Asia. The earth might successfully nourish these hungry billions on a diet of beans and corn and grains, but investors—often multinational corporations—profit more by devoting land and water to the production of livestock for export to a luxury market. Fat cattle, in short, may be starving people to death.

At least one observer believes the cattle trade is not merely bad policy and sham, not merely a colossally wasteful, inefficient, and harmful enterprise. To the popular environmental journalist Jeremy Rifkin it is a "cold evil" that spawns a host of social and moral ills, from ordinary greed and injustice to a disregard for life in general. In the last chapter of his book-length polemic *Beyond Beef,* he writes, "The modern cattle complex represents a new kind of malevolent force in the modern world." He sees this dark force as flowing from "Enlightenment principles," an "evil born of rationalized methods of discourse, scientific objectivity, mechanistic reductionism, utilitarianism, and market efficiency."[5] We confront here a familiar villain: the ideological infrastructure of old-fashioned, left-brained patriarchal capitalism, energized by the optimistic mythology of the American frontier, which operates now as a blind faith in technological progress. One may presume that Rifkin and like-minded partisans would include the mining and logging industries as coagents of this cold evil, with their own appalling record of ecological and cultural devastation. In fact all extractive industries—upon which almost all other industries depend—would probably stand guilty as charged, if scientific objectivity, utilitarianism, and market efficiency are the stigma of infamy.

This bleak view does not, however, mark the extreme frontier of pessimism among environmental radicals. The most thoughtful consider the possibility that humans were destroying and degrading the earth long before the Industrial Revolution. Archaeologists and paleoanthropologists have, after all, supplied a fund of evidence to support such a position. In the view of one of the earliest studies of the impact of agriculture on ecosystems, "Civilized man has despoiled most of the lands on which he has lived for so long." The domestication of livestock, in particular, had a definite and documented impact on the vegetation of the Middle East and the Asian steppes. According to René Dubos, "Early men, aided especially by that most useful and most noxious of all animals, the Mediterranean goat, were probably responsible for more deforestation and erosion than all the bulldozers of the Judeo-Christian World."[6] Even in Paleolithic times, hunting cultures may have decimated the population of large wild mammals, especially the wooly mammoth, steppe bison, giant elk, and wild ass, whose bones occur in great numbers around the fireplaces of those ancestors who mastered flint weaponry.[7]

In his book *Green Rage* Christopher Manes cites such information to demonstrate what he sees as the central importance of the new activism, and its daring scope. "The significance of radical environmentalism does not lie in some jaundiced history of environmental philosophy, nor in the dark urge for political power. Rather, it is based on one simple but frightening realization: that our culture is lethal to the ecology that it depends on and has been so for a long time, perhaps from the beginning. The validity of the radical environmentalism movement rises or falls depending on the accuracy of this perception."[8] In several chapters Manes marks this origin at the Neolithic revolution, and the implications of this tragic reading of history are enormous, far beyond the significance of any contemporary political movement. If the culture humans have developed in the last ten or twelve thousand years—including, presumably, writing, stoneworking, weaving fibers, taming goats, transporting via the wheel—is in fact a catastrophic evolutionary event, threatening the devastation of the entire planet, then we face a dire crisis indeed, and an appalling revelation. We must, it would appear, bear the guilt for bringing creation (terrestrial, at least) to the brink of a final holocaust. Our only possible salvation would then surely demand that we dismantle civilization as we know it, as soon as possible. (The subtitle of Manes's book calls for "the unmaking of civilization.") Cast out every vestige of this culture—goats, cows, books, cloth, wagons, the Louvre—for it is a deadly aberration, a grisly, freakish mutation that undermines the wholeness and balance of all the rest of nature.

We could dismiss these propositions as too inexact and confused to warrant serious discussion, but they provide us with an excellent example of the dubious thinking and passionate, apocalyptic sermonizing that often inspire the radical wing of the environmental movement. Dozens of books like Rifkin's and Manes's have appeared in the last quarter century. They muster a body of scholarly lore and statistics to rouse a determined defense of "wild nature" and to mount a merciless attack on the ideology and technology of modern, industrial civilization. Writers like Murray Bookchin, Theodor Roszak, Jerry Mander, William R. Catton Jr., and the Deep Ecologists (Naess, Devall, Sessions) argue, some calmly and others sulphurously, that a great and final revolution in human consciousness and culture must occur; otherwise the oceans and continents of Earth—and every fish, fowl, beast, and creeping thing alive—may suffer terrible, permanent damage. This catastrophe would of course also mark "the end of humanity's tenure on the planet."[9]

Tony Merten probably read literature of this sort and absorbed it more thoroughly than most. The discourse of gloomy ecoprophets also percolates into all manner of secondary material, especially the newsletters and other publications of environmental groups, from the Audubon Society and Sierra Club to Greenpeace and Earth First! Tom Bill was probably accurate in guessing that the rhetoric he encountered, conversing with Tony, derived from the same sources that inspired the material found in FBI files.

Consider a sentence such as this one: *"The Industrial Revolution and its consequences have been a disaster for the human race."* A number of forceful ecophilosophers, including several of those listed above, have expressed very similar sentiments. Some would insert a qualifier, a prefatory "In many ways," or broaden the statement by substituting all nature for mere humanity, or translate it into poetic analogy, as Bill McKibben does in his landmark essay on the irreversible impact of human enterprise on the earth's climate: "We sit astride the world like some military dictator, some evil Papa Doc—we are able to wreak violence with great efficiency and to destroy all that is good and worthwhile, but not to exercise our global power to any real end." Others might more narrowly define technology or capitalism as the great evil. Thus Jerry Mander propounds the following first axiom: "Living as we do now, using the resources we do, following the inherent drives of a community-oriented technological society, we are doomed to fail." The Deep Ecologists Sessions and Devall concur: "Technological society not only alienates humans from the rest of Nature but also alienates humans from themselves and from each other. It necessarily promotes destructive values and goals."[10]

Like most intellectuals, these authors disagree with considerable acrimony among themselves, especially over the ethical intricacies of ecological policy, but they are in accord on the essential point: devouring the natural world to fuel human "progress" is the road to ruin, a kind of communal suicide. Tony said as much to me at the rendezvous and again to Tom Bill in New Mexico. I have uttered or agreed with such statements myself in the heat of discussion, and I would venture to guess that in the two and a half centuries since Rousseau, a great many people have found the triumph of the arts and sciences (and technology, their sleek modern descendant) to be a highly ambiguous boon. More than a few of these doubtful souls have been haunted by a yearning to renounce the whole show for a simpler, earthier, more spontaneous existence, and in recent years this desire to get away from it all has, for a growing number of people, intensified and darkened into an imperative, one surely influenced by caustic visionaries like those quoted above.

The sentence highlighted above is the first line of the Unabomber's manifesto, and in the ensuing seventy-five pages of that document, a reader will find numerous passages that align Theodore Kaczynski with the most zealous advocates of environmental salvation.[11] Kaczynski himself acknowledges the connection.

> The positive ideal that we propose is Nature. That is, WILD nature; those aspects of the functioning of the Earth and its living things that are independent of human management and free of human interference

and control. . . . Nature makes a perfect counter-ideal
to technology for several reasons. Nature (that which
is outside the power of the system) is the opposite of
technology (which seeks to expand indefinitely the
power of the system). Most people will agree that
nature is beautiful; certainly it has tremendous popular appeal. The radical environmentalists ALREADY
hold an ideology that exalts nature and opposes technology.

Like Rousseau and Thoreau, and most of the major thinkers
and prophets of the environmental movement, the Unabomber
suggests primitive man as a redeeming model and advocates
a simple, rural life that fosters independence and self-reliance in a close-knit, tribal community. "Whatever kind of
society may exist after the demise of the industrial system, it
is certain that most people will live close to nature. . . . To
feed themselves they must be peasants or herdsmen or fishermen or hunters, etc. And, generally speaking, local autonomy
should tend to increase." Kaczynski does not carry this fantasy much further. He notes in his cold and abstract way that
these posttechnological tribespeople might very well devise
a "religion based on nature," but his passion is not for redemption and salvation, either personal or communal. He
seems driven entirely by his fury to destroy the evil empire of
the industrial state.

A glorification of homely subsistence has been common
enough since the romantic age, but mostly as a harmless exercise in nostalgia, a poignant contrast to the grime and tra-
vail of the new metropolis. But the old pastoral dream of es-

cape from the smoky cities has been transformed into a new set of fantasies, lifestyles, and scenarios (outdoor adventure, country estates, survival experience) articulated by a range of interest groups, some of them bitterly antagonistic, devoted to the defense or management or multiple use of wild lands.

People no longer approach the "outdoors" or "the country" naively, expecting only a refuge, a place of healing and restoration. They know nature now has complicated designations: she is monument, park, grassland, or wetland; she is refuge, habitat, preserve, or study area; she is setback, buffer zone, greenbelt, or open space. Her devotees (or exploiters) must secure in advance the proper titles, permits, leases, and reservations, or improvise strategies to bypass those requirements. More and more users of our natural resources, from poachers and marijuana growers to sportsmen, campers, and four-wheelers to loggers, miners, ranchers, and power producers, will routinely assert their version of privacy with alarm systems, locked gates, weapons, and walkie-talkies. The wilderness is now a battleground, ideological and actual.

From this perspective, the Unabomber case is not a unique aberration but only one more consistent example of an alarming undercurrent in our environmental crisis. One could cite here the case of Claude Lafayette Dallas, a young self-made ranch hand and trapper who, in 1981, outdrew and shot down two armed Idaho State Fish and Game officers who had found illegal hides and meat at his camp on a remote reach of the Owyhee River. Dallas evaded a massive manhunt for six months and during his subsequent trial became a hero to many backcountry folks, an incarnation of the legendary mountain men and frontier scouts who lived free at any cost. He represents an extravagant, deadly version of the same inarticulate code that inspired my own father in his encounter with the game warden seventy years ago.

Like my father, Dallas had read and admired Jack London, who was also an inspiration to Chris McCandless, another independent young man who struck out for the West and made a tragic effort to survive alone in harsh, remote places. Jon Krakauer's book *Into the Wild* (1996) relates the story of McCandless's relentless pursuit of solitude in the Alaskan bush, which ended with his death (from hunger and possibly inadvertent poisoning) in an abandoned bus. The book became a best-seller and aroused much controversy, for the story seemed to touch a sensitive nerve in the American psyche, one connected to the opposed neural centers of contempt and pity. The tough survivalists thought the young man a fool; others admired his spirit while regretting that an obsession with wild nature could go so terribly wrong.

These tragedies have certain obvious similarities. They all involve men living by themselves in a remote place, not out of necessity but deliberately and with an unusual strength of conviction. They were not simply nature-lovers; they tried to live on or near unspoiled, untamed lands, day in and day out, forswearing all but the most minimal and absolutely necessary contact with the busy, mechanized society that surrounded them. Dallas lived in a tent, Kaczynski in the most primitive of shanties, McCandless in the junked bus. The latter two were estranged from their families, and Dallas had only intermittent contact with his. All three shared a disregard for convention and an admiration for self-reliant, rudimentary technology. (The most macabre of Kaczynski's signatures was a preference for hand-carved wooden parts in his diabolical personnel bombs. Dallas braided his lariats, hand-loaded his cartridges, made and tooled his own spurs and chaps. McCandless confronted an Arctic winter with little more than a sleeping bag and a .22 rifle.)

However sharply these men differ in their character and philosophy, they are united in a fundamental way: each attempted to revive what he perceived as a historic and ideal bond with the natural world, a bond broken and abandoned by a mechanized, urban civilization, and each was willing to risk everything to achieve that goal. Claude Dallas, the terse buckaroo, might have called this bond simply personal freedom—a right to live off the land without interference from others surviving by one's own grit and savvy. As a young suburbanite enamored of the legend of the frontier, Chris McCandless seemed more intent on building an identity as a backwoodsman, setting a challenge for himself, and living out an extraordinary adventure. The record of his short life reveals an utter disregard for the ordinary goals of wealth, power, and fame, or even domestic comfort and stability, and a fervent faith in the restorative and inspirational value of a spare subsistence in the wilderness.

Theodore Kaczynski is by common measure a homicidal psychotic, yet he believed himself to be the avant-garde of a new kind of revolution, one that would destroy the apparatus of science and technology and return mankind to its original sane condition—the primitive commune of hunter-gatherers and peasants. The Unabomber thought he was devising a world perfectly suited to the likes of Dallas and McCandless. He imagined that there were indeed others like himself, "thoughtful, intelligent, and rational" people who would recognize the truth of his ideology and begin the process of educating the public. (Kaczynski, like Lenin, held an elitist view: "History is made by active, determined minorities, not by the majority, which seldom has a clear and consistent idea of what it really wants.") Eventually, the Unabomber dreamed, even ordinary people might see that much of the injustice, corruption, alienation, and desperation of this fallen world could be banished, simply by subduing the demons of science and technology.

That wistful vision is clearly shared by many fervent ecophilosophers, though they might be outraged at any attempt to link their own theories to those of a pathological killer or even to the tragic fates of an outlaw trapper and a reckless youthful adventurer. I think, however, that some revolutionary ecowarriors and prophets of environmental apocalypse have not fully credited the emotional and psychic impact of the ideas they so enthusiastically purvey. These forces are perhaps easier to chart in the affair of Tony and the cows, which has for me an even more disturbing resonance with the theme I have been tracing--the hope of humanity's redeeming itself by learning again how to live harmoniously in wild nature.

Tony's life resembles in some ways the type outlined above. Like Dallas he was a self-reliant loner, a man of independent mind who forsook the marketplace to indulge a passion for desolate and unspoiled places. Like McCandless he lived his convictions openly, asserted his position honestly. And like Kaczynski, Tony believed in striving to implement an ideal, but he was neither secretive nor psychotic in his promotion of it. He held a position of leadership in a community whose values he shared, and those who knew him best in New Mexico describe him as an unusual combination of meteoric energy and reserve, an intense advocate who guarded his privacy.

Until the last week of his life, when the shadow of suspicion fell, he might have been considered an unconventional but otherwise thoughtful and responsible citizen (as his unblemished record and his reporting for jury duty suggest), and his suicide cannot be easily dismissed as the act of a disturbed or dysfunctional mind. He wrote a last letter unusual only in its plain, direct style and almost offhand tone. If we take his words at face value, he simply came to believe his cause was lost, the green earth doomed—and this was knowledge he could not live with.

His decision to "check out" was extreme, but it was not necessarily irrational. Texts like those mentioned above present a documented, detailed, and very dismal picture of the state of the planet, and Tony was surely familiar with such sources. Also, like the rest of us, he could look almost anywhere about him and see signs of pollution, waste, and degradation traceable to human influence. Or a few keystrokes would give him access to a mountain of information on our declining resource base, the vanishing of species, the spread of new and deadly viruses. He knew as well as the next man what powered these ominous trends. The human population grows, and so does its avidity. Headlong development of manufacturing, transport, commerce, and communications— synonymous with converting wilderness into a global village—has been the dominant preoccupation of colonizing nation-states for more than two centuries, and in Vico's view it is the direction of all human history. How, then, could anyone seriously believe that this priority—the conversion of the planet into goods and services meant to enhance the quality of human life—is being displaced by anxiety over global warming or the ozone hole, let alone by a concern for the welfare of wolves, whales, cranes, or rare desert plants?

Such issues were of dreadful import to Tony Merten. In his farewell letter he wrote simply that the near future terrified him because "I see no hope for humanity or the earth in the grip of climate change (global warming, ozone depletion, deforestation, et al.)." This statement has an additional twist in its context. If Tony indeed pulled the trigger on those thirty-four cows, he would have done so as a reformer reduced to desperation by the failure of tamer tactics. A friend of his spoke of Tony as an "over-the-top rationalist," one who figured out what needed to be fixed and then set forth, implacably, to find the tool and complete the job. But he also had a reputation for bluntness, an almost militant openness—perhaps another way of saying he was "the kind of guy who had a real hard time lying." In this case, he may have discovered that he could no longer lie to himself.

One may imagine that a passionate defender of the earth could push himself to an extreme action, a violent deed, only to discover that his effort brought more grief than gratification. The lawmen were, literally, at Tony's door, and his own life was likely to be, henceforth, severely disrupted. He had not the character for deceit or flight, and the alternative of an excruciating public trial might have seemed intolerable. He would surely have realized that the local environmental community could not support him, would distance itself from gunplay. In short, he might have concluded that even the last, desperate hope—guerilla action for redress of wrongs—was illusory and futile, caused suffering without bringing about change, and that anyway there was no stopping the devastation of the earth by a single corrupted and destructive species—his own.

Tony also wrote in his last letter, "I have never been more satisfied with my life." A startling claim, which some would cite as evidence of extreme denial. Yet he might well have viewed his career with a certain grim approval. He had lived according to his ideals and would sacrifice everything for them. He had dealt with the population problem by sterilizing himself, had worked hard to bring about change within the system, before acting directly to rid the range of the aliens that were ravaging it. He made sure that, hopeless or not, his cause would go forward after his death: he made a will bequeathing a substantial part of his estate to the Southwest Center for Biological Diversity, a Tucson group active in terminating or reducing grazing permits through lawsuit or hostile purchase.

He might also have taken some pride in a decision to maintain silence, refusing either to lie or to compromise the environmental cause by claiming responsibility. Yet such satisfaction would likely be transient, no bulwark against the deluge of despair arising from the revelation that nothing can stop the ruin of Earth, that even the best of ecowarriors—free, self-sufficient, consecrated in and dedicated to preserving the wild—is ineffectual and bound to fail before the onslaught of the rapacious juggernaut of technocivilization. And why should failure be inevitable? Perhaps because to strike effectively at the monster requires one to lead a life of evasion, subterfuge, and dishonesty—an impossible choice for those who most esteem wildness and freedom. Or perhaps—and this possibility is surely the nadir, the absolute zero of an ecowarrior's despair—because the pessimists are right: humans are hardwired at the most fundamental DNA level to exercise their extraordinary cunning in a ceaseless effort to dominate and control everything, to ravage and amass in pursuit of ever-new dreams—the old fatal Promethean striving for godhood.

We will never know for certain whether Tony thought along these lines. All his friends I talked to agreed he was a lonely man, and that burden, not always conscious, can exert a dangerous pressure of its own, not to be discounted in this situation. In the end we have only what the man said, wrote, and did. But that record is enough, I think, to raise at least a whisper of the redeeming hope Tony could not find in his life, because his tragedy is both an embodiment of and a challenge to one of the most venerated and influential traditions in American letters, a tradition at the heart of the environmental crisis.

His legacy is much more than a cautionary tale or a convenient anecdote in a revisionist manifesto. I suspect that the pathos, paradox, and mystery of his life and death are uncomfortably pertinent to all who have seriously confronted environmental issues, at any order of magnitude, and especially to those who have savored unforgettable hours deep in a wilderness and dreamed of someday, somehow, making that experience the staple of life, in a community reborn to that end. If nothing else, the case of Tony and the cows forces us to think through again what we have accepted as true and right, to look again, and hard, at the models we are living by.

CHAPTER
FIVE

*We live in an essential and unresolvable tension between
our unity with nature and our dangerous uniqueness.*
—Stephen Jay Gould, *Hen's Teeth and Horse's Toes*

At Ted Kaczynski's trial in Sacramento in 1998, defense lawyers planned to present his crude one-room cabin as evidence
of a deranged mind. Some observers may have picked up a
hint of irony here. Is not the Unabomber a grotesque yet perfectly recognizable shadow of a revered archetype in American literature? His steadfast contempt for the knickknacks of
civilization, his reclusiveness, and his adoration of wild nature make strong parallels with Thoreau, who also, in *Walden,
or Life in the Woods,* praised his predecessors who lived in
caves and wigwams.

Well aware of his own eccentricity (in the eyes of the
good folk of Concord), Thoreau promoted his humble, hand-
made shanty and his scrupulous autonomy as the mark of a
higher, enlightened life. He, too, chose solitude and used it to
draft a sustained refutation of the notions of progress that
characterized his age. His most heartfelt admonition was
"Simplify, simplify," and he was, in the context of his own

time, quite as technophobic as Kaczynski. Thoreau saw no need for railroads, post offices, or newspapers, and argued passionately for the redemptive power of wild nature, the honor and self-respect to be gained in a life eked out of a few bean-rows and a fishpond.

Thoreau's ruminations inaugurated a remarkable tradition in American letters. Among his descendants we must count not only Muir and Leopold, who, with their distinguished ancestor, compose a kind of sacred trinity of nature prophets, but a subsequent long list of prominent writers, thinkers, artists, and activists. After the learned, Brahminic generation of Bernard DeVoto, Joseph Wood Krutch, René Dubos, and Rachel Carson, came a wave of writers whose lives and works resonated even more strongly with the lonely genius of Walden Pond. As a youth the poet Gary Snyder crafted poems inspired by his solitary duty as a fire lookout on a mountaintop, and Edward Abbey wrote of the grandeur and mystery of the southwest desert from the perspective of an isolated and sometimes alienated park ranger. Alone on Tinker Creek, Annie Dillard distilled from her field notes a series of probing and lyrical meditations on the interplay of life-forms, while Gretel Erlich wrote of the private solace and self-knowledge to be gained from a sheepherder's life in Wyoming.

One could perhaps include also Robert Pirsig and William Least Heat Moon, who took the whole continent as their Walden and speculated, cogitated, dreamed, and moped as they drove its lonesome back roads. And to these one might, without too great a stretch, add a horde of recent scribblers, some obscure and local and some on the best-seller lists: all those who take refuge beside lake or peak or gorge, or set off on solo journeys (on foot or by horse, camel, dogsled, jeep, skis, or kayak), wishing "to front only the essential facts of life," "to drive life into a corner"—and then write about it.

This tradition has articulated, in its maturity, some of the most elevated and esoteric of the contrapuntal themes in the grand symphony of North America. From Thoreau onward, and especially in the mold of Muir and Leopold, writers enamored of the wild landscapes of the New World have emphasized their ethical and spiritual dimension. These apostles tramped the forests and mountains in sun and storm; canoed rivers, lakes, and bayous; tracked and observed every sort of wildlife; slept on the ground beside an open fire. And in this elemental apprenticeship they often underwent what amounted to a religious conversion. For them wild nature became an ultimate presence, a temple or shrine—surely the dominant implicit metaphor of this school of writers—wherein dwelt fundamental and transcendent truths and values.

Hence certain passages from Thoreau have served as a primary and guiding inspiration for the environmental movement:

The West of which I speak is but another name for the Wild; and what I have been preparing to say is, that in wildness is the preservation of the World. . . .

When I would recreate myself, I seek the darkest wood, the thickest and most interminable and to the citizen, most dismal, swamp. I enter a swamp as a sacred place, a *Sanctum sanctorum*. There is the strength, the marrow, of Nature. . . . In short, all good things are wild and free.[12]

Thoreau could depend on his audience to note the etymology of "citizen" and be taken aback by his contrasting of civilization and all that is holy. He probably intended also to evoke ironic echoes of Dante and Bunyan, aware that he was suggesting new and highly heretical forms of worship and redemption. At the dawn of the great surge of migration to America and the settling and development of the Western Territories (primarily in pursuit of Abraham's kind of wealth— gold, silver, and cattle), Thoreau was lighting a votive flame to consecrate an exactly opposite inclination. He would cherish above all whatever was *not* yet "subdued by man"—that wildness, the source of priceless insights gained on his solitary walks in the deep woods. The surprising thing is that this lone candle, the work of a shabby, cranky genius dismissed or ignored by most of his contemporaries, became a beacon to six generations of admirers, and has in the last three decades inspired a serious challenge to the great dynamo of American commerce.

The converts to the new faith have done much more than merely revive enthusiasm for the old pagan trinity of earth, sea, and sky. Writers in this tradition studied the cosmology, folklore, and ethnography of tribal hunter-gatherers, especially the record drawn from Native Americans who preserved remnants of the old ways. In these sources they found sanction and fresh inspiration for their own coalescing beliefs, and what they perceived as an ideal model for living in harmony with the natural world. The aboriginal worldview appeared to sanction the equality of species and the mystical union of all life in a common heritage. Many tribes held themselves the descendants or relatives of mythical animals and treated the earth as a communal, sacred home.

Similarly, new generations pursued the study of Eastern religion as a venerable and articulate tradition of respect for all sentient beings, and as yet another model for a community devoted to a simple, unobtrusive lifestyle. Hindus and Buddhists hold the apparatuses of civilization to be ultimately illusory, and their example encouraged Western students to cultivate the inner life by renouncing worldly ambition and display, in order to confront the self in remote and peaceful surroundings. Native American and Far Eastern ideas and practices are cornerstones in the ethical and metaphysical foundation of radical environmentalism, at least in its most expressive literary incarnation.

Even among Christians a movement began to emend the orthodox view, which gave Adam's progeny dominion over all nature and merely instructed them to be fruitful and multiply. Revisionists came forth to provide congenial new interpretations of Scripture, which stressed an obligation to keep God's earth green and diverse, and to cherish all who lived there—not only man and his cattle, but all manner of fish, fowl, and creeping things.

In a stroke of cosmic irony, given the long and often bitter conflict over matters of faith in the nineteenth century,
scientists also made a contribution to the advance of this new
vision, this eclectic blend of ethical and spiritual beliefs cen-
tered in the natural world. Among the more influential of these
writers one might list Albert Schweitzer, Rachel Carson, Loren
Eisely, Gregory Bateson, James Lovelock, and Lynn Margulis.
Carson's book *The Silent Spring* (1962) is often cited as the
spark that began the environmental revolution by rousing the
public to an awareness of the damage done to the ecosystem
(and public health) by toxic pesticides. Schweitzer became a
kind of patron saint for organizations devoted to animal welfare, which heralded the more militant and ideological animal rights movement. Lovelock and Margulis provided a
simple, scientific model for more explicitly spiritual notions
of nature. Their Gaia hypothesis (1975) gave environmentalists a goddess to revere and a convenient symbol for the essential unity of all living things.

By the early 1980s this various company of essayists, natu-
ralists, wanderers, poets, philosopher-prophets, scientists,
scholars, and journalists had begun to approach the critical
mass of an active, visionary faith, which, while too diverse
and unorganized to be called a coherent religion and too intellectual and idealistic to survive as a major political force,
yet showed itself capable of inspiring a holy zeal and intensity of conviction among its followers. This evangelical fervor and militant energy stemmed from the perception—already emergent in Thoreau—that the great cathedral of na-

ture was being systematically violated, turned into fodder for a ravenous industrial civilization. And this civilization was clearly approaching its peak of vigor and virulence in America at the twilight of the twentieth century. Powerful interests were at work there, driving wells into the sea floor, laying huge pipelines across arctic wilderness, constructing nuclear and coal-fired power plants, logging the last stands of old-growth forest, and damming the mighty rivers of the West.

In response, the fiery prophets of the new, green faith began to call openly for revolution, the liberation of nature, the disintegration of contemporary American society and culture. This revolution, as envisaged by the most far-seeing ecological philosophers, was not simply a drive for political power or autonomy but a radical and complete transformation of the dominant moral paradigms governing human behavior, both social and individual. The aim was to change, fundamentally, the way people looked at themselves and the world, to substitute the teachings of Black Elk and Sakyamuni for those of the Old Testament and Adam Smith. Or in Judi Bari's characteristically succinct and secular formulation (speaking as leader of a major faction of Earth First!), "We want to have an entirely new society that's based on achieving a stable state with nature instead of exploiting the earth."[13]

Sympathetic philosophers and cultural critics present the ecowarriors and their supporters as the last incarnation of a grand tradition of moral activism, the heirs of the abolitionists, suffragists, Industrial Workers of the World, Civil Rights volunteers, and peace marchers. They are seen as bearing a torch whose rays illumine the path of justice and freedom, reaching now, for the first time, beyond mere humanity. They struggle to protect the rights and autonomy of wild creatures, to guarantee the integrity and sustainability of the biosphere— the entire intricate, interdependent web of soil, waterways, air, and plant and animal communities—as the only way to provide salvation for all, an alternative to mankind's fatal, narrow obsession with self-aggrandizement. Assigning such primacy to nature, according to one noted ecophilosopher, is henceforth humanity's "highest and noblest moral calling." Or as the chief historian of the tradition phrases it, the new environmental ethics proposes "arguably the most dramatic expansion of morality in the course of human thought."[14]

Creating an "entirely new society" to embody the noblest of human aims—a utopia—is of course one of mankind's enduring and fabulous dreams. But it is a dream seldom realized, and it carries the risk of metamorphosis into nightmare, a possibility that seems to increase in direct proportion to the fervor of its partisans, the scale of the dream, and the speed with which the faithful try to make it come true. Since the radical environmental movement is a response to what is seen as a global crisis, now at an advanced and accelerating stage, the rhetoric of its visionaries has taken on a certain stridency. Time has almost run out and we have everything to lose, they postulate, so whatever the risk and however small our numbers, we represent the last, best hope for humankind and so must launch our crusade—now. Or in the hortatory, figurative style of Dave Foreman, when he was Earth First!'s most visible spokesperson: "It's time for a warrior society to rise up out of the earth and throw itself in front of the juggernaut of destruction, to be antibodies against the human pox that's ravaging this precious beautiful planet."[15]

This kind of trumpet blast illustrates well enough the perils of striving to realize Ecotopia. The true believers in this tradition commonly draw spiritual nourishment from a private, direct experience in a wilderness, as did the forerunners Thoreau, Muir, and Leopold. (Foreman's epiphany came on a trek into Mexico's forbidding Pincate Desert.) That experience is deep, powerful, life-changing—essentially a kind of divine revelation, and a new convert wishes, naturally enough, to sustain and make permanent the conditions that foster salvation. Therefore the temple of wild nature must be kept inviolate, its boundaries extended, and more faithful sanctified at its altar. It is after all the foundation of life itself, a wellspring of hope, the place—be it grove or canyon or sea-cliff—of peace, regeneration, transcendence.

Unfortunately, this noble agenda collides immediately with the "juggernaut of destruction." That mighty engine turns out to be pretty much the whole of modern Western civilization, which some argue is now essentially identical with American high-tech entrepreneurial capitalism and the global, consumer society it is feverishly building ("McWorld" in one critic's phrase[16]). In such a confrontation the attitudes of peaceful contemplation and moral admonition have little immediate impact, and give way soon enough to the firebrand's rhetoric of revolution. After all, a handful of the faithful find themselves posed against the most formidable force in human history—the leading industrial nations, their colossal productive capacity harnessed to ever-faster systems of manufacture, transport, and communication—the whole driving incessantly to transform the raw materials of nature into human artifacts, real or virtual.

So a passionate and righteous but vastly outnumbered minority must struggle to avoid being overwhelmed or ignored, and their lofty language must be translated into deeds. Faced with such a daunting challenge, the visionary vanguard often chooses to act in a highly dramatic and symbolic way, so as to capture the public's imagination or at least its attention. Strategies range from the phantom saboteur to the martyr who casts himself in front of the bulldozer, from the mass demonstration, with costumes and banners, to the lone hunger strike or outrageous graffiti, like the three-hundred-foot black plastic "crack" Earth First! hurled across Glen Canyon Dam.

A rough general rule suggests how aggressive such an action will likely be: the direr the threat, the more virtuous and inclusive the alternative order foreseen, the easier it is to justify disruptive or violent tactics. Or (as the vanguard would style it), the more implacable the tyranny, the grislier the genocide, the more potent and shocking the antidote needed. Hence those recurrent and uncomfortable paradoxes: war in pursuit of peace; murder to protect embryonic life; burning villages to liberate them. Or returning to Tony and the cows, how would we explain, in terms of mankind's "highest and noblest moral calling," the methodical act of blowing out the brains of two-day-old calves, their mothers already on the ground kicking and bleeding to death? And how explain further, in the same exalted language, a bullet through the brain of an active, thoughtful man with half his life to live?

The committed ecophilosophers might well suggest that shooting thirty-four head of cattle, however foolish tactically, is ideologically not without justification and perhaps even a plus in the long run, assuming the public has to be jolted into an awareness of the misuse and degradation of public land. If the beef industry is in fact a "malevolent force in the modern world," responsible for imperiling whole watersheds and their wild inhabitants, as well as starving the earth's impoverished multitudes, and if a hopelessly hamburger-addicted public is indifferent to these consequences, then it is not difficult to imagine how an individual, especially one of keen conscience and passionate disposition, might be motivated to strike directly at such a menace. In the light of many of the texts considered here, such action would constitute a blow intended to defend a sacred cause—the restoration of that wildness that alone can assure the preservation of the world.

The problem is that the act of shooting somebody else's cows, for whatever noble purpose, violates another code of great moral authority. Inspired by the Enlightenment political philosophers, the first citizens of the United States were devout believers in their freedom to secure a livelihood, acquire property, and trade and travel at will, without fear of interference or confiscation, as long as they did not impede the like liberty of others. The wilderness, or rather its riches, belonged to whoever got there first, though no man was supposed to take more than he could reasonably use, and in the end all would benefit in the distribution of new wealth. The owners and leasers of what is left of that wilderness now invoke the same ethic and pretend to extract natural resources for the public good, creating jobs for workers and supplying goods to consumers. Miners, loggers, and ranchers often appeal to this frontier tradition and claim the right—or as they often view it, the obligation—to develop new sources of wealth from the people's lands, for the general welfare. They are, they argue, simply meeting the demand for wood products, fuel, minerals, and food. Nowadays, of course, they also hasten to guarantee that this extraction will be sustainable indefinitely, without significant loss to the environment.

The ranchers' long-term leasing of public rangeland poses some especially thorny and explosive issues. Mining and logging companies often have substantial private holdings, and turnover is common among their lower-echelon employees, who migrate from one work site to another. At the worst a gang of dedicated environmentalists could cost them a few paychecks, but logging shows and veins of ore run out anyway, and periodic unemployment is a given in these trades. Ranchers, on the other hand, think of their grazing permits as an integral, indispensable, and permanent base for their livelihood and lifestyle. They spend a good share of their days driving a pickup or riding a horse over this range, which they are likely to know more intimately than anyone else and which they also profess to love—perhaps less audibly than the environmentalists but with equal sincerity.

It is not uncommon for a ranching family to have held the same ground for three or four generations, and their reputation and identity are inseparable from their spread, which almost always includes what they view as their rightful, traditional lease of public land. A reform that terminated this arrangement in order to preserve habitat and rare species could mean economic ruin and the end of a proud tradition for thousands of families. It is doubtless also alarming to consider the possibility that impatient zealots might launch such reform on their own, with a rifle. However misguided their assumption of a right to public land might be, one can surely understand the ranchers' current campaign to organize, lobby, and fight back on all fronts.

The scrap between the ranchers and the environmentalists develops from a clash of moral visions, and such struggles are likely to be intense, complicated, and exhausting. It is always difficult to chart a middle course when strong feelings are in play and the stakes are high, but this conflict has even deeper and trickier currents than most. It may represent a fault line in the national soul and an example of that "unresolvable tension" Stephen Jay Gould finds between nature and human nature. Talk of compromise, balance, trade-off, or mitigation by any number of spokespersons, facilitators, politicians, and editors may be irrelevant and empty.

There may in fact *be* no bridging of the fundamental chasm between the advocates of the sacred wild, the believers in a new order valuing the pristine ecosystem above human advantage, and loyal supporters of MAC (modern American civilization), this huge, complex system whose economic life is an ever-expanding cycle of production and consumption. Their opposition could be permanent, not because cowboys and wolf-lovers are so bitterly antagonistic but because the populace whose allegiance is supposedly the prize in the dispute is itself profoundly and utterly ambivalent, has already a strong attachment to wilderness (at least as an ideal) while its survival is inextricably bound to the industrial juggernaut, which in our context means hamburgers and sheetrock. This paradox is exactly what unites the partisans on both sides, though that is a truth too uncomfortable to face.

I am concerned here with a human casualty, because I think Tony Merten was caught in the vise of this paradox. I would like to find and learn the lesson of his story, which may be instructive to others who have grown uncomfortable with their position on the environmental crisis, who suspect that we have not thought well enough nor far enough nor honestly enough about our real place in this world, and have allowed a comfortable righteousness in the service of certain wishful fantasies to mold our convictions.

I have traced my own unease to an obvious and fundamental source—the archetype posed by the sage of Walden Pond: a man alone in the wild, attaining through hardship and simplicity a kind of enlightenment, a virtuousness—including the original sense of manly power—that strikes through the pretense and artifice of society to a superior truth. Thoreau elaborates that archetype throughout his work but perhaps most forcefully in the final pages of *Walden*. He says he learned by his experiment how, for one who pursues his dreams with confidence, "new, universal, and more liberal laws will begin to establish themselves around and within him . . . and he will live with the license of a higher order of being. In proportion as he simplifies his life, the laws of the universe will appear less complex, and solitude will not be solitude, nor poverty poverty, nor weakness weakness."[17]

From this vantage point, one need not depend on family or the marketplace or a well-ordered society. Such preoccupations appear secondary, if not trivial and absurd. The great triumphs are interior and private, the hard-won discoveries of the lone explorer who journeys simultaneously into the wilderness and the unknown inner continent of the self, returning as a superior being, perhaps poor and in rags like the Hindu holy ones, but radiating the potential of a new destiny for humanity. Hence Thoreau claims he wished merely to stroll through the woods with "the Builder of the universe" and "not to live in this restless, nervous, bustling, trivial Nineteenth Century, but stand or sit thoughtfully while it goes by."[18]

Of course there is good evidence that this resolve did not endure. (The fact that Thoreau *wrote* is itself a refutation of such detachment.) He admits to no lack of human intercourse even during his retreat to Walden, confessing that upon occasion twenty or thirty guests descended on his cabin for the evening. And no incorrigible recluse would write, as he does, "I think that I love society as much as most, and am ready enough to fasten myself like a blood-sucker for the time to any full-blooded man that comes in my way."[19] We might also recall that after two years on the pond Thoreau returned to his old life: the lecture circuit, writing articles, and cadging meals and odd jobs from the Emersons.

Yet his wish to stand apart and alone is not just a pose or self-delusion. I think many of us understand Thoreau intuitively and immediately and forgive him any inconsistency because we also have known a powerful yearning for solitude and rapt contemplation in some secret glade or cove. That is one good reason why *Walden* continues to move readers profoundly. Given the actual circumstances of our lives— the traffic, paperwork, bad air, garbage, crime—we are charmed and heartened by the myth of a tough individualist who steps away from the mad rush of progress into one of nature's wild sanctuaries and there achieves a "higher order of being"—all at the expense of a borrowed axe and a few pennies a day.

Most of us ultimately smile, sigh, and shrug off the temptation to think seriously about severing our connection to society and plunging into the wild to test this fantasy, or we settle for a tepid substitute, a time-share Walden for weekend retreats. But very rarely someone like Tony Merten makes a serious effort to revive the myth. Tony retired from the rat race early and built his home on a dirt road twenty miles deep in an empty desert. He lived there by himself, as self-sufficient as possible. He devoted his time to communing with nature or reading, thinking, and writing about how to preserve wild places and creatures. To a few friends he showed a notebook uncannily like Thoreau's journal—an exact and careful accounting of his domestic economy—and spoke with pride of the frugality that would allow him to survive indefinitely without involvement in a wasteful, destructive culture.

But the upshot of his commitment was not peace or contentment or revelation, as we usually understand these. It is doubtful that the laws of the universe appeared to him any less complex. For one thing a new century has come and gone, and it was no nervous, bustling, trivial thing, but a roaring Behemoth that consumed Waldens by the tens of thousands, swallowed forests, gouged out mountains, drank rivers dry, and moved a whole population of farmers, fishermen, drovers, trappers, hunters, and woodcutters into factories and offices. For another, the Thoreauvian archetype is singularly vulnerable in the face of such an onslaught: a contemporary hermit-philosopher who would approach nature as a personal shrine designed for *solitary* worship is likely to face an agonizing spiritual trial.

For radical descendants of Thoreau, solitude is practically out of the question. Instead they find themselves under immense pressure to transform their private spiritual quest into a moral crusade. Appalled equally at the ravening desecration of all things wild and free and at the skill of image makers in selling this horror as responsible, progressive management, they run a grave risk: bitterness and outrage may tempt them to make of their enthusiasm for nature an automatic mark of moral superiority and, simultaneously, to rationalize any involvement with Mammon as bondage forced upon them by the ubiquitous evil of technology.

These tendencies are widely apparent in the radical environmental movement, with what I believe are most unfortunate consequences. Put bluntly, this is the situation: a righteousness tinged with unconscious hypocrisy poses perhaps the single greatest obstacle to the movement's realization of Ecotopia. The purer the green idealist and the more he or she demonizes technocivilization (on Web sites, for example), the more contorted his or her conscience becomes.

Various Deep Ecologists and ecophilosophers encourage the faithful to think of themselves as the leading edge of the great, emancipatory liberal tradition, fighting for a last, sweeping extension of the most basic rights. Yet both thinkers and activists are well aware of a certain isolation: they are after all a tiny minority undertaking the complete regeneration of human culture. It would be no surprise if such an elite but embattled avant-garde came to view itself as the original, purest, and highest ideal of humanity, and the rest of their species as a toxic muddle of ignorance and corruption. Something like this position emerges in Foreman's contrast between the "warrior society" and the "human pox." A similar contrast occurs in the writings of some ecofeminists, who trace the continuing destruction of the natural world to the egotism and greed of a patriarchy tricked out in the false morality of rational humanism, which is in their view a late and perverse usurpation of a primal, creative, nurturing, holistic, integrative, and above all *natural* order shaped by the female spirit. The environmentalists' near worship of traditional Native American thought and their ringing denunciation of corporate, technological capitalism constitute the same classic polarity: the spiritual versus the worldly, the noble versus the base, the saved versus the damned.

Such loaded comparisons are likely to offend the good citizens who are the very object of the environmental crusade, at least in a democracy. When they enjoy relative prosperity and share a common faith in, or at least acceptance of, the system that produced it, citizens will not greet reformers with enthusiasm and may react with hostility to the argument that their successes are in fact mortal sins. Nor does it improve matters to portray ecoradicals as inheritors of the grand liberal tradition of equal rights and opportunities. Securing the freedom and humane treatment of all living things or defending the integrity of the entire biosystem will not be widely perceived as an equivalent of preventing the slavery, oppression, and slaughter of human beings. The public might agree that clubbing seals and shooting children are both immoral acts, but they will not see them as tragedies of the same order of magnitude. To treat them as such is to invite outrage or ridicule, and for readily understandable reasons.

In the political struggles of the past, the oppressed could speak for themselves, and they argued from the powerful position that they were, except for the fabricated distinctions of rank and privilege, quite like their oppressors. They were in fact agitating to escape being "treated like animals," to force recognition of the very obvious fact that they were completely human, not property to be owned and traded at someone else's pleasure. From the concession of common identity and shared ancestry, a system of reciprocal respect and just distribution was bound to grow. The doctrine of natural rights, as delivered to us from previous centuries, is rooted firmly in the principle of the common kinship of all humanity. Equality and liberty are alike born from fraternity.

In contrast, the new ecological missionaries present the human species as a deluded and deadly colossus tyrannizing the entire biosphere, and simultaneously they volunteer themselves as enlightened advocates to speak on behalf of voiceless hordes of wild creatures and their vanishing habitat. To be purified and redeemed, the new sermon continues, we must abandon our oppression of others and extend justice to all the earth's innocent and wronged (wolves, whales, wombats, watersheds). This noble aim can be accomplished by preserving and expanding wilderness areas, and by rediscovering and following the simpler and sounder path of the ancients, the primal peoples who respected and preserved the natural world for so many millennia. To the enlightened few, no higher imperative exists for the human race; for if we cling to our obsession with technology and our consumerist addictions, we—and perhaps the planet—are doomed.

This prophecy is the culmination and logical conclusion of the Thoreauvian tradition. (Though one may doubt that its founder could abide some of its more bombastic and reckless extensions.) It calls for renunciation and self-discipline, for a simpler and more rustic culture, for a society dedicated to the wild and free, where the ascetic and spiritual (or in some scenarios the feminine) will be ascendant. In this moral transformation, the lust for dominance, narrow ambition, greed, and chauvinism will be exposed as deadly infections of the soul. Humankind will be, at last, at home in an egalitarian universe.

Few ideals in the contemporary scene have such thrilling sweep and moral altitude, or demand so great a leap of faith. The lofty grandeur of this vision, however, entails some troublesome side effects. For starters, the new religion of nature can inflict on its converts a great psychological burden, which bears an unsettling resemblance to that imposed by the old, anthropocentric Puritan code. Scientific rationalism (knowledge is power) becomes original sin, technology its fruit, and history the saga of the progressive seduction of the human race by this corrupt and malign influence. An inescapable conclusion: MAC is irredeemably foul, and most of humanity is consumed by its temptations and therefore damned.

If we accept the evidence (which exists in good measure) that our civilization is dangerously wasteful, destructive, insensitive, and short-sighted in its pursuit of dubious material ends, and agree further that our ravaging of wild lands and our extinction of whole species are both sign and consequence of these failings, and finally concede that this disregard for other forms of life is in some sense immoral, then we face an excruciating dilemma. Either we acknowledge a collective guilt of monstrous magnitude—our own species become the Great Satan—and dedicate our lives to a spiritual rebirth as hunter-gatherers, or we abandon ourselves to heretical compromise and rationalization; we backslide to the position (articulated or not) that humans must devise universal moral codes to their own advantage or at least fabricate a system of special dispensation and allowance to provide for the extraordinary complexity of their needs. (This takes the form, usually, of a strenuous effort to believe in the myth of a feasible balance between economic growth and the environment.)

Both these alternatives can lead the bewildered pilgrim even deeper into the treacherous bog of environmental ethics. Standing aloof from technological civilization in practice is quite a different thing from rejecting it in principle, and attempting to live by current ecological theory may lead one into untenable, if not absurd, postures. Paul Taylor, a distinguished academic ecologist, has emphasized that simple justice will require humans to "locate and construct their buildings, highways, airports, and harbors with the good of other species in mind." J. Baird Callicott takes up the issue of managing human behavior generally and argues for a stern, ascetic kind of socialism, inspired by the pre-Neolithic lifestyle. He believes humans should "reaffirm our participation in nature by accepting life as it is given without a sugar coating," like the cavemen for whom "the chase was relished with its dangers, rigors, and hardships as well as its rewards . . . a tolerance for pain was cultivated . . . population was routinely optimized by sexual continency, abortion, infanticide, and stylized warfare." While Callicott grants (regretfully) that we cannot hope to duplicate "the symbiotic relationship of Stone Age man to the natural environment," he believes the "ethos" of our distant ancestors "could be abstracted and integrated with a future human culture seeking a viable and mutually beneficial relationship with nature."[20]

More recently Jack Turner's book *The Abstract Wild* called for a renewal of the ancient ways, a "new tradition of the wild" that would synthesize the very elements I identify as central to the radical environmentalist creed. "Such a tradition of the wild did exist; it is as old as the Pleistocene. Before Neolithic times, human beings were always living in, traveling through, and using lands we now call wilderness; they knew it intimately, they usually respected it, they often cared for it. It is the tradition of the people that populated all of the wilderness of North America, a tradition that influenced Taoism and Hinduism. . . . It is the tradition that emerged again with Emerson and Thoreau." Then, in a startling coda, Turner qualifies his position to allow for a certain amount of intercourse with the opposed traditions of industrial civilization—a version of the balance so dear in political greenspeak. "I am convinced such a life is still possible. I love my Powerbook, my Goretex gear, and my plastic kayak. But I also make a point to eat fritillaria, morels, berries, fish, and elk. I want to feed directly from my place, to incorporate it. When I die, I wish my friends could present my body as a gift to the flora and fauna of my home, Grand Teton National Park, because I want my world to incorporate me."[21]

Many of us have reveled in fantasies of living as cave-dwellers or aborigines (inspired in part by popular writers from Chateaubriand through Rice Burroughs to Auel), and many of us hunt, fish, and gather partly to make a point—a kind of homage to the simplicity and freedom we imagine our ancestors enjoyed. A very few of us are even perhaps serious when we say we would like to live off the land all the time and would prefer the rude, sometimes hazardous lot of the fisher, hunter, or primitive farmer over a career with IBM or Yale University. But surely we can recognize that most of the time circumstances do not permit the realization of such fantasies, and in fact conspire to discourage them. The residents of Manhattan or Los Angeles could not handily feed from their place, unless they turned cannibal; nor would it be a simple matter even to offer up the urban deceased to become one with gulls, pigeons, or alley cats. And of those who do migrate to the wilderness, only rare eccentrics have managed, in their embrace of the primitive, to regress further than the horse-drawn cart and black-powder musket.

For most of us the only feasible course is, again, compromise—weekend camping, the occasional check to Greenpeace, regrets over the effluent of our corporate employers. That is of course where we have been for some time: the status quo, which isn't really working, given global warming, the ozone hole, vanishing species, and so forth. Hence the introspective and conscience-ridden are attracted—or driven—to Thoreau's model of a grand, simple, personal revolution.

Unfortunately, taken out of the context of a roundtable discussion among ecophilosophers bubbling with speculation and theory, propositions like those cited above are preposterous. No ordinary citizen would take seriously a proposal to locate highways and airports according to the interests of ground squirrels and geese (who would instantly recommend moving them off-planet). Nor, citing Neolithic precedents, would he or she petition Congress to legalize infanticide or "stylized warfare" as population control measures. It is equally hard to establish a claim of blood-kinship with ancient subsistence hunters when one is warmed by Goretex, travels in a four-wheel-drive SUV, makes a kill with a high-velocity rifle, and then hauls out the Powerbook to email the news home.

What astonishes in these scenarios is not their absurdity but the tenacity with which even highly intelligent, well-informed scholars hold to the vision of a reborn humanity living a simplified life in harmony with wild nature. They are by no means alone in this romantic attachment. The potent myths of Eden, the golden age, and the noble savage are apparently immortal; they represent a force like gravity, penetrating layers of asphalt, concrete, asbestos, vinyl, sheetrock, and acrylic pile carpeting, and operating as well beneath any number of elaborate disclaimers, qualifiers, and masks of tough-minded realism. Allied with our national myth of the limitless frontier, this complex of values and feelings associated with nature has made almost every American an environmentalist of some sort: harried commuters, carpenters, ranchers, students, stockbrokers, salespeople, corporate executives, professors of philosophy—all are likely to vouchsafe support for some version of the wild, to exert considerable effort to at least drive through a natural landscape a few times annually.

In this context the passionate work of the radical new ecomoralists is especially informative and stimulating. The vibrant field of environmental ethics forms a fascinating mosaic of contradiction and revelation, confession and evasion, insight and denial. For in striving to configure a powerful, ancient dream to our modern, technological age, these thinkers have exposed universal dilemmas and dangers. In following their assumptions to logical extremes, they hold up a mirror that shows even more than they intend: not only the ominous features of a global technological culture but also the elaborate shifts in our own psyche that obscure our real connection to and dependence on that culture; not only the spiritual regeneration wild nature offers but also the risks of a zealous and single-minded faith in such transcendence. If nothing else—and by itself this would be an invaluable contribution—the struggle to forge codes of environmental ethics makes us keenly aware of how complicated, bizarre, and treacherous the human mind can be, even at its soundest and sharpest, and may bring us to suspect an appalling discontinuity between our radiant, inspiring dreams of nature— the forests, jungles, tundras, and deserts of Earth at the dawn of our history—and the facts of our current lives, chief among them the incessant drive to order and contain this world, make of it a reified ideology—organized, understood, "respected."

The question is, of course, which vision is the true and accurate one. Gaia can be seen as everywhere vulnerable and ailing, a reduced and fragmented presence despite the apparent grandeur of certain protected monuments; or she may reveal a sinister, Shiva-like face, the signs of an impending apocalyptic collapse—drought, flood, earthquake, plague, and blight; or her most devout worshipers may persuade us that she will finally begin to mend and recover her original glory— but only when most of the recent "achievements" of human culture have been eradicated and we (in greatly reduced numbers) have returned to the caves and huts of our ancestors; or finally, the believers in compromise and balance may prevail and redefine the goddess as a kind of suburbanized grande dame of exotic pagan lineage who--with proper management—will remain our generous, efficient, and supportive partner in the great old enterprise of "getting on."

This last scenario is of course the one we are currently trying to live out. In countless advisory councils and consensus groups environmentalists sit down with resource extractors, government agents, and scientific consultants to hammer out the policy of balance; or lawyers representing such interests struggle to adjust those policies to a new bias. The result is demoralizing to those immersed in the Thoreauvian tradition, because one does not strike a new balance with the sacred or adjust the parameters of essential and ultimate value. From the radical environmentalists' perspective, the direction of this compromise turns out to be, whether the shift is fast or slow, almost always in one direction: away from the wild and free, toward an increasingly high-tech, micromanaged, humanized world.

I have a hunch my friend Tony pursued a train of thought

very like this one, influenced by texts such as those I have cited. And I see well enough how the process could lead one to an ultimate despair. Outside his door the genetically engineered meat machines supplying McDonald's and Burger King were trampling rare desert habitat. Whatever he—or anyone else—did to stop them was never enough. He also knew perfectly well that this local issue was only a microcosm of a planetwide disaster, the continuing ruin of forests and grassland and rivers and the air and sea. In short, an Earth so damaged and defiled, so tilted toward compromise, that no hope for the return of sacred wildness could survive.

For the most impassioned of the latter-day descendants of Thoreau, there is no longer a refuge and vantage point, a Walden from which one can contemplate the arrival of yet another "restless, nervous, bustling, trivial" century. If Gaia is a dying and doomed goddess, and humanity is responsible, then only two options remain: a hopeless, if heroic, struggle against one's own species, or suicide. Stated so baldly, such thinking may appear pathological. But for those who hold ecological integrity to be the highest and holiest ethical principle, and also believe that most of humanity is fatally driven to destroy that integrity, the conclusion is merely logical.

I have the temerity to advance these speculations and suppositions because I am no stranger to the sort of despair that I imagine between the lines of Tony's curt final letter. In the course of this inquiry I came to recognize that I had felt in my own bones, for years, the roil of a very disturbing emotional vortex underlying my love of wilderness and my fretting over the direction and momentum of American culture—a force like the tug of a distant, dark star whose elements include grief, rage, and shame.

This hidden complex became a chronic feature of my own interaction with the environment. Call it, in the initial stage, just a bummer, a swift little depression of the kind that sprouts, simple and immediate as a mushroom, when I jet along five miles above the patchwork of clearcuts and contouring roads that the Sierras, Cascades, Siskiyous, Bitterroots, and Rockies have become; or when, commuting to work as one drop in a great freeway river, I am startled to see that the last strip of orchard has disappeared into a new trailer park or mini-mall, and so it will be all concrete and glass and asphalt and neon and flags and balloons and searchlights now, in a California valley where kids once picked oranges and caught catfish under oak trees. Or—the mushroom erupting suddenly to monstrous, rainforest size—when, at the halfway point in a long journey to a remote wilderness, I slump before a TV in a motel, brain open and ravaged by a lurid torrent of imagery, every pixel calculated, targeted, charged with lust, fear, hope, envy, greed, joy, pity, hatred—a fast-food menu of human emotions devised to stir new addictions in a consumer—and I realize that I scarcely have the will to turn away, because there is also a fascination, an ugly little thrill here, a deep-rooted connection to this virtual world and its nightmare levels of speed, power, aggression, and desire. And even if we do, in an access of guilt, punch the remote, we know in our hearts that the motel bed and bath, the burger and shake, and the interstate are also involved with that world, are conjured by that same hit—the juice, the rush, the jazz of American culture—which has become now a necessary, an automatic part of everything, including this wilderness experience. We realize, in short, the full import of Pogo's famous proverb: We have met the enemy and he is us.

I had only to imagine these uneasy moments extended and deepened into a habit of mind in order to sense the immense pressure this kind of environment could exert on a sensitive intelligence devoted to preserving wild nature, sacred nature. The votary would see all too clearly how rapidly and inexorably the wild earth—what is left of it—is being consumed, transformed, sold as mere space, or at best tricked out as an exhibit, and how the overwhelming majority not only accept but celebrate this development. The shrewdest of these partisans rise to fortune and high office; the world is their oyster, a metaphor with a new appalling accuracy. And millions—billions—of people seem eager to ape them, yearning for a chance to revel in the feast.

The gastronomic conceit can be extended. The American dream is, after all, a piece of the pie, and the pie in question is not primarily a game refuge or a view of the mountains or even a guarantee of no more than a few parts per million of heavy metals in our drinking water. The pie is a big comfortable stylish home full of labor-saving and entertaining appliances; a powerful and luxurious vehicle; unlimited access to colorful, exotic other lands or, nowadays, exciting cyberworlds wherein the mind may coast free of the ordinary constraints of space-time; and also of course a handsome stylish mate and healthy, intelligent, well-adjusted children already slated for lucrative professional careers, hence a family that glories in sharing all these triumphs and sweet diversions, so that eventually in that long sunset-time of fly-fishing and golf, world-cruising and fine dining one can look back with pride and know an ultimate peace and satisfaction: know that one has done well and done the right thing, and all in this world is as it should be.

Nature (as separate from natural resources) supplies at best a sprinkle of spice over the crust of this grand American pie. It becomes recreational space, a campover or course. The slope or pipe or run. A trout stream or float river. Great elk country or the best place to backpack or bike. To be sure, millions of Americans are passionate lovers of outdoor sports, and many of them claim that they would do nothing else if they could, that they "live" only to fish or ski or kayak or climb. But that claim is empty for all but the very few who smuggle their passion into a career, becoming guides, packers, commercial fishermen, game biologists, or writers for outdoor magazines. One doubts that even as many as one in a thousand among the general public would permanently trade the fruits of civilization—the house, car, pool, PC, and TV— for a shack in the wilderness with only a fishpond and a beanfield to sustain life. And of those, only a much smaller number could actually survive on such terms.

Hence the wilderness functions more and more in a dual role: as research station and religious theme park. Experts visit it to gather data; tourists come to feel awed and reverent, to recharge their spiritual batteries. The patches of great forest, rugged coast, and trackless desert that remain to us are indeed becoming like shrines—old shrines carefully preserved and museum-like, sometimes with ticket booths and convenience stores at the gate. The visitor seeks an experience: sighting a rare species (or catching and tagging it), or testing a new theory of migration, or enjoying a reassurance that the old tales are true—trees can be so big, canyons so deep—and that we *do* care about nature.

This model of memorial wilderness amounts to a final horror for those who yearn for a completely new, biocentric society dedicated to fostering the wild and free. In its simplest formulation, that horror emerges as the consequence of a certain perspective on history and evolutionary theory. From this angle, one sees clearly that *Homo sapiens* has become, over fifty millennia, the chief instrument of rapid change in the planetwide ecosystem, that so far human settlements and enterprises (from the spear and the goat to the atom bomb and Angus) have not succeeded in shaping this earth to enhance diversity or to ensure the stability and integrity of ecosystems, and that free will seems to operate primarily by subduing wild nature to the service of certain human obsessions.

One may conclude that humans have evolved into creatures driven to make the world always new—or else God has taken traits of the ant, beaver, packrat, and coyote and joined them with a scrap of his own nature to form this hairless, dream-driven ape. In any case, here is a species fated to divert rivers, to cut down or burn off forests, to bear seeds into new climes, to domesticate useful animals, to invent tools and weapons of greater and greater efficiency, to multiply and exterminate competing or inconvenient species, to erect temples, to connect oceans, to write symphonies and epics, to create laws, to travel faster than speeding bullets, to yearn after the stars, and finally, of course, to explain themselves and what they have done. In this view we are always imposing our vision on the raw stuff of reality. In our ugly, frank, contemporary lingo, we might say that humanity is programmed to fashion ever-new, virtual habitats from this given earth, and the advance of each new package of brain software entails the obsolescence of an older one.

This conclusion can have a dramatic impact on some temperaments. The fervent apostles of the arriving information age (a new worldwide web of life) may be exhilarated, confident in their belief that science and technology are indeed the deepest and finest articulation of our unique gifts, our best hope of solving whatever problems present themselves. For true believers in this camp, wild nature still has immense value as a data bank and control group; it is a big laboratory, necessary to the advancement of knowledge and (with the right precautions) a refreshing and exhilarating place to work. It is seen as an important element of our stake in the future. Respecting this resource does not, however, by any means imply that humanity should refrain from controlling and directing it. In fact modern scientific conservationists will argue that to be sustained now a wilderness *must* be monitored and managed, for it is no longer a vigorous, dominant environmental context but a set of fragmented and endangered microsystems. The subtlest of these thinkers even argue that nature (the term henceforth forever between invisible quotation marks) is already virtual, an imaginative human construct defined by various laws, reports, surveys, statistics, and of course projected desires and expectations, all of which change over time. Summarizing a semester-long conference of distinguished scholars on the topic of reinventing nature, its chairman asserted that "'nature' is not nearly so natural as it seems. Instead, it is a profoundly human construction."[22]

The majority of humankind would rally to this standard, in its more pragmatic and mundane formulations, for these reassure us that it is our destiny to remodel the world to improve our lot. A currently fashionable version presents the image of corporations, consumers, and wise government in a new green coalition, girding to meet the challenge of improving both the quality of life (jobs, health care, recreation) and the environment (air, water, soil, habitat). Pollution credits are bartered to enforce overall public health standards without interfering unduly with production goals; manufacturers get tax incentives for recycling or developing sustainable energy sources; developers protect sensitive wetlands in exchange for permits to build on other "less significant" or already degraded areas; user fees fund research into game populations and the creation of interactive preserves; and—as ranchers in the Great Basin are already demonstrating—the wild buckaroos can be retrained to the new ecotainment industry, guiding dudes on river floats or elk hunts or cattle drives styled as a summer vacation for families. In this vision nature is always, and "naturally," undergoing dynamic transformation at the hands of her programmers, who have at last learned that to prosper they must know her thoroughly, respect her limitations, and manipulate her with great care.

For a small but articulate and influential minority, such propositions are an evil, Manichaean betrayal of a higher reality and represent a mortal danger to the earth and all her inhabitants. For these dissenters, nothing damns humankind more absolutely than this anthropocentrism, this hubris rooted in the Greco-Judeo-Christian tradition. The assumption that man is the pinnacle of creation and center of the universe, empowered to subordinate and exploit all other life-forms, is thus the ultimate original sin, the single, stark root of our environmental crisis, with its malaise and alienation, its hectic, constrained, overstimulated, and inequitable lifestyles, and its augury of mass suffering to come. To claim further that such domination and exploitation are the natural functions of our species is blasphemy; for in the faith of the most biocentric ecophilosophers, the ethics of this planet authorize no species to assert dominion over all others, and "the killing of a wildflower . . . is just as much a wrong . . . as the killing of a human."[23]

The sharp division between these two views of nature has the force and seriousness of a religious schism, with the attendant risk of fanaticism. A step further and the conclusion may be that those who wilfully crush a rare lily are murderers. Two steps, and humans may be compared not to flowers but to thistles—a rude, overabundant species in need of severe thinning. On the other hand, the most committed advocates of unrestrained technology and free enterprise contemplate manipulating and controlling the genetic stock of humans and, presumably, most other living things. And from this position it is not a great stride to foresee a kind of preemptive strike against all "undesirable traits" of poorly adapting "naturals" in order to favor the survival of a brave new world, a society of elite clones and the slave-species selected to serve them. To partisans of the wild such gradual, ghoulish engineering would be as horrible as old-fashioned genocide.

We are some distance yet from a jihad involving such extreme ideologies, but there is little doubt that conflict over environmental issues will intensify under the twin pressures of population and aspiration. It also seems likely that much of this conflict will involve public lands—those lonely, semi-arid basins and ranges where the cattle roam. The mosaic of these western lands contains most of what remains of wild America: the last uncut forests, unplowed river bottoms, and unsettled valleys—a national treasure (from another perspective) of natural resources—and the whole region is now experiencing rapid development, especially as a site for recreation and retirement communities.

What happened with Tony and the cows in Luna County is, as I see it, a kind of alarm directed to all the parties in the range, timber, and water wars already underway in the West. So-called wilderness areas could well be the scene of even more arson, bombings, gunplay, and vandalism in the upcoming century, as both resources and living space grow scarcer. To forestall such moves, moderate environmentalists and the extractive industries will doubtless be driven to devise painful compromises, leaving the radical activists isolated as usual. But compromises are inherently unstable, and such deals will by no means guarantee salvation from either economic or environmental collapse.

What Thoreau meant by wildness is mortally endangered by this scenario, which presents worshipers of pristine nature with the same Hobson's choice that Tony confronted. It is very possible that, barring a miraculous mass conversion or a catastrophic die-off of humans and their institutions, there is no hope for such wildness and therefore no salvation of the world, at least in the sense Henry David intended. Of course many of us want to believe otherwise, want it with a sometimes desperate ferocity; for the vision of a past primeval wilderness grows more luminous and compelling in direct proportion to the conversion of its remnants into mere monuments. We recognize that much more than scenery is at stake, that we are also losing an ancient and powerful perspective, one that invests mountains and rivers with a spiritual (and sometimes terrifying) grandeur and a promise of ultimate harmony and meaning. Yet holding to this vision with unrelenting fervor, in defiance of everyday facts, can exact a very high price. That is the tragic implication of Tony's story.

The Luna County affair has no villains or heroes, and is far too mysterious and ambiguous to serve as a clear example of anything, except perhaps a dark turbulence in our nation's soul, the strange blend of greed and grief that has accompanied our assault on those purple mountains and fruited plains that once stretched from sea to shining sea. But the case does suggest that our environmental crisis is not to be resolved simply by more study, larger data banks, an improved flow of information, or the exercise of rational choices by educated and aware citizens. Facts and ideas are not the only vital elements in this crisis. We are imbedded in or disjoined from nature in many complex and contrary ways, and these bonds and barriers derive, ultimately, from a configuration of very deep and powerful emotions peculiar to humanity. We know—or should know—that this perverse pattern of our own nature drives us to acts of astonishing creativity, which can also serve a hideous destructiveness, and at the same time draws us toward a passive, solitary, and mystical worship of the living wild world.

That we are simultaneously angel, demon, and beast is not news. But our hybrid species has become dangerously schizophrenic since the Industrial Revolution. Various factions are furiously at work dismantling, managing, and loving this world to death. They all claim to serve the public interest, and each can cite venerable texts and hallowed myths to justify its efforts. Each seems to believe the struggle is a moral one, a question of good and evil. As the friction between these different versions of environmental consciousness intensifies, the partisans lose sight of a simple, central truth: this conflict is inseparable from our humanity. *We* are the source of this whole problem, of all the good and evil, and none of us is without a generous share of both.

I have criticized radical environmentalists for a failure to acknowledge this common responsibility and have attributed the movement's narrowness and zealotry, at least in part, to the influence of Thoreau, as he appears in *Walden*. But Thoreau's genius was not a model of consistency, was in fact and most appropriately, wild. In the midst of his sojourn at Walden Pond, Thoreau made an excursion to Maine, and in the upper reaches of the Penobscot he experienced at last "a wholly uninhabited wilderness, stretching to Canada." His reaction to that exposure, as well as his later and better-known hardship climbing Monadnock, shows us a different view of nature, less comfortable in its assumptions, darker in its implications, and very pertinent to this discussion. Standing on the bank of a wild river, looking north into a dark and trackless wood, the sardonic Concord intellectual, a slashing critic of civilization and its mores, fell silent. The last traces of human enterprise—the skid tracks of the loggers' oxen—had faded away, and except for an occasional transient hunter or Indian, this wilderness was absolute. Thoreau was alone, one may infer, as he had never been before, face to face with himself. The comment he made then sums up perfectly the point I am after, the point I want environmentalists of every stripe to examine deeply. "Here, then, one could no longer accuse institutions and society, but must front the true source of evil."[24]

One might stop here and let the shadow of that sentence lengthen (and I did so in earlier drafts). But confronting ourselves can be a beginning as well as an end, and once we choose to go on we must take a direction, which is perhaps what hope is, in its barest and most elementary state. From Tony's life we might learn that "wildness" is no longer so pat and perfect a spiritual destination, but his death—a legend now in the scrubby flats and shark-fin ranges of Luna County—has also delivered some lessons.

The man who replaced Tony as chairman of the local Sierra Club was, ironically, a fourth-generation rancher named Jim Winder. When Tony was implicated in the cattle-killings and ended his own life, Winder says he had to "take a big step back and rethink the whole thing." Originally he had allied with the environmentalists because he recognized that urban expansion was their common enemy. But the tensions developing in the modern range wars—underscored by Tony's suicide—brought focus and urgency to his preoccupation. After chairing the Sierra Club for a term Winder launched a fresh campaign to save his ranch and way of life. He co-founded a new environmental group, the Quivira Coalition, devoted to studying ways of managing livestock operations with an eye to watershed protection and habitat restoration. A series of workshops and conferences recruited support for these ideas, including media attention and grants from state agencies. Within a year Winder was accepting an award from the New Mexico chapter of the Wildlife Society, appearing on radio shows, and giving lectures out of state.

His other move was to break up his 52,000-acre Heritage Ranch by creating a limited number of homesteads of from 50 to 150 acres, with conservation easements that would prohibit any further subdivision and at the same time allow ecologically sound cattle operations to continue on surrounding land. Winder produced a detailed brochure and video for the project, stressing its guarantees of open space, seclusion, and wildlife enhancement, and in less than a year sold more than half the homesteads to buyers from "all over the United States." This enterprising rancher told me candidly that he would never have sold any part of Heritage Ranch willingly, but he could see that he might lose it all if he didn't. Like stock-raisers throughout the West, he faces hostile pressure from two sides: environmentalists and recreationists who want to stop or severely restrict grazing on public land, and local government officials eager to urbanize the landscape and boost taxes. As Winder saw it, his choice was between giving in to uninhibited and terminal development or finding some creative, preemptive compromise that would allow him to stay in his trade.

So the Quivira Coalition organizes field projects to demonstrate the benefits of managing range land to reduce erosion and restore riparian corridors. They encourage new systems of intensive grazing, rapid rotation, and long recovery, and tout the use of mobile electric fencing and of windmills or solar pumps to move water to off-channel stock tanks or ponds. Scientists, bureaucrats, and environmentalists are invited to review and evaluate the results. The argument is that cowboys can also think green, since a healthy ecosystem—good grass and fresh water—produces better beef. To Winder, the whole enterprise is not so much idealism as a straightforward strategy for survival in the modern West.

In the video advertising Heritage Ranch, Winder delivers his last appeal on horseback, overlooking the rolling range and blue, glittering *ciénagas* that are a prime selling point for the property. He has outlined the deal straightforwardly: water, power, and telephone service are available, yet every homestead will be tastefully secluded from the others. Fishing, hunting, camping, and even—if one is interested—a little cowpunching are featured attractions. With easements written into the deeds, no further development can occur. A kind of southwest Walden for a very reasonable thousand dollars an acre, and absolute security for a rare and wonderful place. To dramatize this final message, Winder nudges the horse into a modest gallop away from the camera toward the sunset as the final credits come up. He looks good in the saddle, like a real cowman, and it is beautiful country. But I would guess he feels just a bit foolish on this ride. He isn't going anywhere. There are no cows on the hills ahead, no sign of any habitation, no other rider. As soon as the camera stopped he most likely pulled up and cantered back to ask, with a one-sided smile, if it was a good shot. That is, would it help save the ranch?

A good question for us all. We may wonder these days whither we are galloping, what we are escaping from, and what preparing, with our burgeoning, lusty population and its plans for electric autos, recycled plastics, pollution credits, and getaway homesteads. Are we honest realists, taking responsibility for the excesses of our own nature? Or is it all about beauty, fresh air, and privacy for those who can afford them, here on Earth Ranch? Should we settle for the best deal we can get, or are we at heart true believers who will spur away from a troubled future to reinhabit our ancient pastoral dream?

Tony finally turned away in despair from such conundrums and stalked into the darkness. Winder is finagling to merge his cattle operation into a wildlife refuge and retirement community, a conjuration that probably leaves no time for speculating on such nuances. Two men with very different visions and the same goal—to protect a beloved land. One a success story and one a tragedy, I suppose, though if Babe were still around he would be wondering right about here whether there was maybe a little more to it than that.

Maybe the best we can do is recognize the courage and sincerity of this odd pair in wrestling with issues that seem to be getting more excruciating all the time, and bear away a lesson or two for ourselves. First, that the primary uncharted and untamed wildness lives in our own hearts, and the salvation of the world may indeed depend upon coming to terms with that revelation. And second, that if we are to save ourselves and this little atom we inhabit, we must do so together. Our heroes had better not kill anybody—including themselves—or we are lost.

1. Will C. Barnes, *Western Grazing Grounds and Forest Ranges* (Chicago: Sanders Publishing, 1913), 16; Wesley Calef, *Private Grazing and Public Lands: Studies of the Local Management of the Taylor Grazing Act* (Chicago: University of Chicago Press, 1960), 57; Council for Agricultural Science and Technology (CAST), Task Force Report no. 129, *Grazing on Public Lands* (N.p.: Ames, Iowa, 1996), 15.

2. Quoted in the *Las Cruces (New Mexico) Sun-News,* March 2, 1996.

3. Statistics from Jeremy Rifkin's *Beyond Beef* (New York: E. P. Dutton, 1992); *The World Almanac and Book of Facts* (Mahwah, N.J.: Funk and Wagnalls, 1996); and CAST, *Grazing on Public Lands.*

4. John McPhee, *Irons in the Fire* (New York: Farrar, Straus, and Giroux, 1997), 199.

5. Rifkin, *Beyond Beef,* 283.

6. Vernon Carter and Tom Dale, *Topsoil and Civilization* (Norman, Okla.: University of Oklahoma Press, 1955), 6; René Dubos, *A God Within* (New York: Scribner's, 1973), 160.

7. Marvin Harris, *Cannibals and Kings* (New York: Vintage, 1977), 30.

8. Christopher Manes, *Green Rage: Radical Environmentalism and the Unmaking of Civilization* (Boston: Little, Brown, 1990), 22.

9. Murray Bookchin, *The Ecology of Freedom* (Palo Alto, Calif.: Cheshire Books, 1982), 18.

10. Bill McKibben, *The End of Nature* (New York: Penguin, 1990), 78; Jerry Mander, *In the Absence of the Sacred* (San Francisco: Sierra Club Books, 1991), 382; Bill Devall and George Sessions, *Deep Ecology* (Salt Lake City: G. M. Smith, 1985), 48.

11. All Kaczynski quotations are from the Unabomber's manifesto;

complete text in the *Washington Post,* September 19, 1995 (special supplement).

12. Henry David Thoreau, "Walking," *The Portable Thoreau* (New York: Viking, 1947), 609, 613.

13. Oral presentation for workshop at the Trinity River Rendezvous, July 2, 1995.

14. Paul W. Taylor, "Are Humans Superior to Animals and Plants?" *Environmental Ethics* 6 (summer 1984): 160; Roderick Nash, *The Rights of Nature* (Madison: University of Wisconsin Press, 1989), 7.

15. Foreman quoted in Manes, *Green Rage,* 84.

16. Benjamin R. Barber, *Jihad vs. McWorld* (New York: Ballantine, 1995), 4.

17. Thoreau, "Walking," 562-63.

18. Ibid., 568.

19. Ibid., 390.

20. Paul W. Taylor, *Respect for Nature* (Princeton, N.J., 1986), 270. J. Baird Callicott, "Animal Liberation: A Triangular Affair," *The Animal Rights/Environmental Ethics Debate*, SUNY Press (Albany, 1992), 56.

21. Jack Turner, *The Abstract Wild* (Tucson, 1996), 90-91.

22. William Cronon, "The Trouble With Wilderness; or, Getting Back to the Wrong Nature," *Uncommon Ground* (New York, 1995), 25.

23. Paul W. Taylor, "In Defense of Biocentrism," *Environmental Ethics* 5 (Fall, 1983), 243.

24. Thoreau, 89.

About the Author

Will Baker's books include *Backward: An Essay on Indians, Time, and Photography*; *Mountain Blood* (essays); and the novels *Track of the Giant*, *Shadow Hunter*, and *The Raven Bride*. He has written articles on logging, game poachers, the global teenager, and hunting in the Peruvian Amazon. He operates a small farm in Northern California, and rides on local cattle drives.